The Mines And Its Wonders

by

W.H.G. Kingston

Double 9 BOOKS

The mines and its wonders

by W.H.G. Kingston

ISBN: 978-93-64282-96-3

Published by

DOUBLE 9 BOOKS

2/13-B, Ansari Road
Daryaganj, New Delhi – 110002
info@double9books.com
www.double9books.com
Tel. 011-40042856

ABOUT THE AUTHOR

William Henry Giles Kingston (1814-1880) was an influential English writer, best known for his adventure novels targeted at young readers. His works, particularly those with nautical themes, have captivated audiences with tales of heroism, exploration, and moral integrity. Debut: Kingston's literary journey began with the publication of "The Circassian Chief" in 1844. Genre: He specialized in seafaring adventure stories that were highly popular in the Victorian era. Output: Prolific in his writing, Kingston authored over 130 books, many focusing on nautical adventures and aimed at young readers. "Peter the Whaler"(1851): One of Kingston's early successes, detailing the adventures of a young whaler.

"The Three Midshipmen"(1873): Part of a series chronicling the exploits of British naval officers.

"The Three Admirals" (1891): Continuation of the naval adventure series, showcasing themes of bravery and exploration. Impact on Children's Literature: Kingston's adventure stories have had a lasting impact on children's literature, particularly in the adventure genre. Kingston's contributions to literature have made him a celebrated author, particularly known for his ability to inspire and entertain with stories of exploration and heroism. His works remain a testament to the adventurous spirit of the 19th century and continue to be enjoyed by readers around the world. Enduring Popularity: His tales of the high seas and distant lands continue to be appreciated for their timeless appeal and adventurous spirit. Kingston's contributions to literature have made him a celebrated author, particularly known for his ability to inspire and entertain with stories of exploration and heroism. His works remain a testament to the adventurous spirit of the 19th century and continue to be enjoyed by readers around the world.

CONTENTS

Chapter One
The Miner's Dangers

A hum of human voices rose from a village in the centre of England, but they were those of women, girls, and children, the latter playing in the street, running, skipping, laughing, singing, and shouting in shrill tones, the former in their yards or in front of their dwellings, following such avocations as could be carried on out of doors on that warm summer evening. Not a man or lad, not even a boy above eight years old, was to be seen. On one side of the village far away could be distinguished green fields, picturesque hills, widespreading trees, and a sparkling stream flowing in their midst; on the other, nearer at hand, a dreary black region, the ground covered with calcined heaps, the roads composed of coal dust or ashes, and beyond, tall chimneys sending forth dense volumes of smoke, which, wreathing upwards, formed a dark canopy over the scene. Then there were large uncouth buildings, above which huge beams appeared, lifting alternately their ends with ceaseless motion, now up, now down, engaged evidently in some Titanic operation, while all the time proceeding from that direction were heard groans, and shrieks, and whistlings, and wailings, and the sound of rushing water, and the rattling and rumbling of tram or railway waggons rushing at rapid speed across the country, some loaded with huge lumps of glittering coal, others returning to be refilled at the pit's mouth. Those high buildings contained the steam-engines which worked the machinery employed in the coal mine; the tall chimneys carried up the smoke from the furnaces and produced the current of air which kept them blazing. The deafening noises came from cranks, pulleys, gins, whimsays, and other contrivances for lifting the coal from the bottom of the mine, pumping out the water, loading the waggons, ventilating the shafts and galleries, and for performing duties innumerable of various descriptions. As the evening drew on, the women retired into their cottages to prepare supper for their husbands and sons, whose return home they were now expecting. Already the corves which took them down to their work in the early morning must be on their way up to the surface, and it is time to have the savoury messes ready for dishing up. Abundance is on the board, for the miner's wages are sufficient to supply him with what would be luxuries to an ordinary labourer above ground; but were they far higher, could they repay him for a life of constant danger, of hard incessant toil, and the deprivation for more than half the year of a sight of the blue sky, the warming rays of the sun, and the pure air of heaven, except on the one blessed day of the week when he enjoys them with the rest of God's creatures? For months together he descends the shaft in the gloom of morning and does not return till darkness has again shrouded the earth.

Many of the good wives had looked at their clocks to judge when to take off the bubbling saucepans from the blazing fires, when, to their dismay, they felt the earth tremble beneath their feet, while a dull rumbling sound like the discharge of musketry struck their ears, coming from the direction of the works. Pale with terror, they rushed out-of-doors to see a vast black mass of dust and smoke rising into the air and forming an inverted cone, beneath which, for an instant, could be distinguished shattered beams and planks, corves and pieces of machinery, which quickly fell again to the earth. The next instant a darkness, like that of early twilight, pervaded the atmosphere, and fine ashes, such as are ejected from a volcano, fell in a thick shower to the ground, which it covered to such a depth that the feet of the terror-stricken women left their imprints on it as they ran towards the scene of the catastrophe—some shrieking and lamenting, but, in most cases, the intensity of their alarm preventing them from giving utterance to their feelings. Among them a young woman, superior to the rest in appearance, went hurrying on towards the pit's mouth, her hand held by a little boy, who had evidently grasped it, refusing to be left behind, when startled by the explosion, she had quitted her cottage. Her fair hair, escaping from beneath her cap, streamed in the wind; her countenance exhibited the most intense anxiety. Her boy, among the oldest of those who had remained that morning in the village, was well able to comprehend what had occurred, yet he did not cry or shriek out, but did his utmost to keep pace with the woman's rapid steps.

"Perhaps father and Mat had come up before the blast happened, mother," said the boy in a hopeful tone. "They would be stopping to see how things are going on, or maybe to help any poor fellows left in the pit." The woman answered only by a gasp. "Don't give way, mother dear," continued the boy. "We shall find them both well above ground, depend on't." Still the woman made no reply; her heart told her that her worst anticipations would be realised. She and the rest of the women from the village arrived in a short time at the pit's mouth, where, among the ruined buildings, the broken machinery, and the heaps of rubbish, they rushed frantically here and there seeking for the bread-winners of their families, many uttering piteous wails when they sought in vain for their loved ones; while others, when they were discovered, bursting into shrieks of hysterical laughter, as they flung their arms round the men's necks, led them off to their homes. Some of the miners had, it appeared, come up just before the explosion; but what was the fate of the rest, far beyond a hundred in number, still below? Some, it was surmised, might have escaped death, and many brave volunteers came forward ready to descend to their rescue. All was quiet—the shaft appeared to be free—a fresh corve or teek was procured—a rope attached to the gin, to the shaft of which a party of men putting their shoulders worked it with the strength of horses. The corve descended with its adventurous crew down the shaft. The young woman with the little boy had been among those who had sought in vain for a husband and son. "Have any of you seen John Gilbart and his boy Mat?" she asked of those who had come out of the pit and of others standing by. No one could give her any information about her husband, though one had replied that he had seen young Gilbart leaving the trap at which he had been stationed.

Unlike the other women, on hearing this she uttered no cry, but stood speechless and trembling as near as she could venture to the pit's mouth, where she waited, with intense anxiety, the return of the corve to the surface. "Don't take on so, mother dear," said little Mark, who felt her hand trembling. "They say some may have escaped, and things may have been worse above than they were down at the bottom. Perhaps they threw themselves flat on their faces, and let the blast pass over them. I heard father say, only the other day, that was the best thing to do when fire-damp breaks out. He wouldn't have forgotten that, mother, would he?"

"I pray Heaven that he did not," she answered in a scarcely audible voice. Minute after minute went by, while the brave explorers who had gone below were searching for their comrades. How that poor mother's heart ached as she thought of what had too probably happened to those she loved. Night had come on, but torches and lanterns and a blazing fire not far off lighted up the scene, casting a lurid glare on the dark figures of the men, the lighter-coloured dresses and pale faces of the women, and the surrounding ruins. At last the cry arose that the corve was ascending. The eager crowd pressing forward could with difficulty be restrained from impeding the men working at the gin. Then came the shout, "They're alive! they're alive!" and six dark figures stepped out on the ground. They were soon recognised by their wives or mothers, and hurriedly dragged off to their homes, while the rest of the women, bitterly disappointed, waited till the basket should again come to the surface. The same scene was again enacted, and the rescued now reported that there were more to follow, though how many they could not tell.

Little Mark and his mother waited with trembling hearts. Those they longed to see had not appeared, and to their anxious inquiries no satisfactory reply was given. Neither John Gilbart nor his son had been seen. At length, another party came up from the depths, but this time there were five boys borne in the arms of stronger men. Alas! two were motionless—the arms and heads of the others drooped helplessly down. The poor mothers pressed forward—Mark and Mrs Gilbart among them. "That's Mat—that's Mat!" cried the child, as one of the first was placed on the ground. The mother, kneeling by the side of the boy, gazed into his face. Too truly she recognised her son, but no responsive glance came from his once bright eyes. "Oh, speak to me—speak to me, Mat," she exclaimed. There was no reply. She took his hand, it was icy cold. Then she knew that her boy was dead. The doctor came. "I grieve for you, my poor woman; he is past recovery," he said, and went on to attend to others. Little Mark sat by his dead brother's side, gazing at him with awe. No one disturbed him. Mrs Gilbart waited on, hope not yet abandoned. More men came up, some fearfully injured. They reported that the rest were in the workings far away—already the mine was on fire, the heat and smoke unbearable—that it was a miracle any had escaped, that all but themselves must have perished. Heartrending were the wailings and shrieks and moanings which arose at this announcement, confirmed by the viewer and overmen. Still many lingered on in the hopes that the corve might be again sent down, but the viewer forbade any to descend, as it must prove their destruction. At length some men came to carry young Gilbart's corpse to his mother's cottage. She and Mark followed with tottering steps. The sad truth had forced itself on her that she was a widow—the two bread-winners of her household gone. Still it was some poor consolation to have recovered the body of her son. Many had not that—they were destined never again to see those they loved. More explosions took place, and the report was spread that the whole mine was destroyed.

This was, however, not the case. Science enabled the manager to triumph over the fiery element raging below. By completely closing the mouths of the shafts, the atmospheric air was excluded, and the flames extinguished. After nearly three months' labour, the mine was explored, and the bodies of the dead, scorched and dried to mummies, were recovered. None could be recognised, and they were buried in a common grave. Mrs Gilbart knew that her husband was among them. The pit was again opened. Fresh labourers arrived from other parts, and once more those dark galleries became the scene of active industry. The cottages were required by the fresh comers, and Mrs Gilbart, with her son and her little girl Mary, a year younger than Mark, would have been compelled to go forth houseless and penniless into the cold world, had not an uncle of her late husband, a hewer at a pit a few miles away, offered to receive her and her children into his house. She thankfully went, hoping to maintain herself and others by her needle.

Simon Hayes had been a miner from his boyhood. Though there were some soft places in his heart, he was rough and untutored, and he had many of the faults common among men of his class. He had a wife much like himself in several respects, but he had no children. Though receiving good wages, he had saved nothing, having spent them extravagantly in obtaining luxuries for himself and his wife, for which they cared but little. By refraining from these, he was well able to feed these additional mouths, and for some time his wife made no complaint at his doing so. Still there was nothing saved up for a rainy day. Simon Hayes took mightily to little Mary. There was nothing he thought too good for her; but he showed no affection for Mark. He was a boy doomed to labour as he had been, and the only labour he could think of for him was down in the mine, first as a trapper, then as a putter, and finally as a hewer. Mrs Gilbart shuddered when he alluded to the subject. She had hoped to bring him up to some trade which he could follow above ground, though it would be several years before he would be old enough to be apprenticed. "But he is not very strong, and he is my only one, uncle, you know," she answered. "Let him go to school first. I have taught him what I could, but he will get on with his learning there faster than at home."

"What's the use of learning to a miner?" exclaimed Simon with a gruff laugh. "However, you must have your way, Mary, and I don't mind paying for his schooling, though, look ye, if times get bad, he'll have to earn his bread like the rest of us." Mrs Gilbart thanked her uncle, hoping that the evil day was put off for a long time. Little Mark went to school, and being fond of his books, made rapid progress in reading and writing. He thus soon possessed himself of the key of knowledge. Little Mary was also sent to a girls' school, and being bright and intelligent, soon became a favourite pupil of the mistress. At length Mrs Hayes fell ill, and her niece's time was so fully occupied in attending on her, that she could gain nothing by her work. Then there was the doctor to pay. Simon also was laid up for some weeks from a severe bruise by a fall of coal. "I can't stand this no longer, niece," he said one day. "The next time I go down the pit I must take Mark with me." Mrs Gilbart begged hard that her boy might remain above ground. She would take him from school and try to get employment for him on a farm. Simon was obdurate; if she would not agree to his wishes she might leave his house. Her fears were all nonsense, the boy would do well enough in the pit, he would get tenpence a-day as a trapper—on a farm he couldn't get twopence. Without telling her what he was about to do, the first morning he returned to work he took Mark by the arm and led him along to the pit's mouth. He had brought a flannel suit. He made the boy put it on. "Now, Mark, we are going into the pit, and you'll do what I tell you when we get down," he said, as if it was a matter of course. "I've arranged with the manager to take you on from to-day as a trapper. Though you may not like it at first, you'll soon get accustomed to the work, and so let's have no nonsense. Here's the corve all ready to go down—come along."

Chapter Two
Learning to Watch

Simon, taking Mark by the hand, stepped on to an iron frame-work or cage, suspended over the pit's mouth. "Take hold of this bar and don't move as you value your life, boy," he said.

Mark obeyed. Several other men and two boys stepped on to the cage, it began to descend. Though little Mark had been hearing of mines all his life, and felt no especial unwillingness, yet all seemed strange about him. It appeared to him by the dim light of the lamps which his uncle and the other men held in their hands, that the shafts were rushing upwards at a fearful rate, while the light of day, which he could still see above him, grew gradually less and less. A giddiness overtook him. He might have fallen, had not his uncle still held him by the shoulder. How long he had been descending he could not tell, when he found the cage come to a stand-still, and that he was down beneath the surface of the earth, a thousand feet or more.

The rumbling of the trains of laden waggons coming to the shafts, the faint voices of the men in the distance, were the only sounds heard, while the lights which flitted here and there only served to make the long vaulted galleries appear more gloomy and dark.

"Come along, Mark!" said his uncle, shouldering his pick and spade, and holding his lantern before him.

As they stepped out of the cage, they found themselves in a gloomy vault, on one side of which a huge furnace was unceasingly roaring, while at the other were the stables in which a number of horses, mules, and donkeys were kept. Before them was the main gallery, about eight feet high and the same wide, arched over with bricks four thick, and extending three miles away from the mouth of the pit. Out of it for its whole length opened shorter galleries or side galleries where the coals were now being won. In all of them rails were laid down for the waggons to run on, and on each side were seams of coal, in some places narrow near the top, in others close to the ground, and in some there was coal from the top to the bottom. At the entrance of these side galleries were doors which had generally to be kept shut, and were only opened when the waggons, loaded with coal or returning empty, had to pass through. After Simon and Mark had proceeded a couple of miles along the main gallery, they stopped at one of these doors. "This is to be your post, Mark," said Simon.

"When you hear the waggon coming, you are to open the door, and as soon as it is passed to shut it. Mind you don't go to sleep. You'll be in the dark, but that won't hurt you, and if you feel anything running by, you'll know it's only a rat. It won't touch you while you are awake. I began my life in this way, so must you. There, go and sit down in that hole cut out for you. When you hear the rolley coming, pull that rope, which will open the door. There, now, you know what to do. Take care that you do it," and Simon, leaving his nephew, proceeded on to the farther end of the working. He then commenced operations on a new cutting which the under-viewer had marked out for him in the side of the gallery. It was about three yards square, and was to be about four feet six inches back under the bed of coal, he began by hewing away about two feet six inches from the ground and working upwards, cutting out the coal with his pick, shovelling it into a large corve or basket which stood at hand ready for the reception of the lumps. At first the work was tolerably easy, as he could stand upright and swing his pick with all his force. As he got deeper and deeper into the bed, he had to fix a strut or post with a cross beam to support the weight of the roof, and he had to get the coal out by stooping down low or resting on his knees. Finally he had to work lying down on one elbow, swinging his pick over his head with the other arm in a way a miner alone could have used it.

Occasionally the boy called the putter came by, shoving a rolley or little band-waggon before him. On to this the full corve was lifted and the empty one left in its place. Sometimes he proceeded by cutting a space on each side of the square bed of coal, from the roof to the floor. He then bored a hole in the middle of the block, into which he rammed a charge of gunpowder, and having lighted it by a slow match, retired to a distance. The powder exploding, shattered the whole mass, and it came tumbling down to the ground in fragments. This could only have been done where no foul air was present, otherwise the moment the lamp was opened there would have been a fearful explosion, and he, with many others, would perhaps have been killed. He laboured on incessantly until dinner time, when he and all the men in the working, including the putters, came out, and taking Mark with them, repaired to a central spot where there were casks of water, and seats, the only accommodation required by the rough miners. Here their dinners, which had been sent down during the morning, were eaten.

"Well, how do you get on?" asked his uncle of Mark.

"I kept awake, opened the door when the rolleys came by, and shut it again after they had passed!" answered Mark.

"That's what I had to do!" said Simon.

"I only wish that I had a candle, and had brought a book down to read. I should not have minded it much then, although it was a hard matter to keep awake!"

"You were not afraid, then?" asked another man.

"What was there to be afraid of?" asked Mark. "I heard noises, but I knew what they were, so I did not mind them!"

"You'll do!" said his uncle in an approving tone. Mark ate his dinner, and then went back to his trap. He there sat all alone in the dark, anxiously waiting for "kenner" time. It came at last, and Mark heard the words "kenner, kenner," which had been shouted down the pit's mouth, passed along the galleries. It was the signal for the miners to knock off work, and return to the upper world.

Mark, however, could not venture to move until his uncle came for him. He was very thankful when he saw the glimmer of a light along the gallery. Slowly it approached. It was carried by his uncle, who having closed the door, led him along through the main tunnel towards the shaft. Together they ascended, and returned home. Mrs Gilbart had been dreadfully alarmed at her son's absence, until told by a neighbour that she had seen him going along with his uncle towards the pit's mouth.

A mother's eye alone could have recognised him, so greatly changed was he by the coal dust. She soon, however, got that washed off, and dressed him again in his clean clothes. He did not complain or ask his mother to keep him out of the mine, so, although still with an unwilling heart, she allowed his uncle to take him. The next Saturday he received five shillings, which was as much as she could make by stitching all day, and sometimes late into the night, by her needle. Simon was well pleased with Mark, and reported, after he had been some weeks at work, that no fault had ever been found with him. He was always awake, and ready to open and close his trap at the proper time. When a little bigger, he would become a "putter," and have the employment of rolling the waggons along the tramways.

Coal mines, it should be understood, are worked in various ways, some in squares, or what is called the panel system. The main roads are like the frame of a window, the passages like the wood-work dividing the panes of glass, and the masses of coal which at first remain, may be represented by the panes themselves. After the various passages have been cut out, the masses are again cut into, pillars only remaining, each of which is about twelve feet by twenty-four feet in thickness. At length these pillars are removed, and props of wood placed instead, and thus the whole mine is worked out. There are miles and miles of passages in which tramways are laid down, leading to the shaft, up which the coal is raised. As the air in the mine has a tendency to get foul and close, it is necessary to send currents of wind into the passages to blow it away. The chief object is to make the wind come down one shaft, and then to bring it along through the passages, and so up by another shaft. If the wind which came down were allowed to wander about, it would produce no good effect. The traps or doors, such as the one at which Mark was stationed, are used to stop it from going through some passages and make it move along others until the bad air is blown out of them. To create a powerful current, a large furnace is placed at the bottom of one of the shafts, which is called the up-cast shaft, and the foul air is cast up it. Often, notwithstanding this, the heat below is very great, and the hewer working away with his heavy pick is bathed in perspiration. Where no bad gas is generated, open lights may be used, but this cannot often be done with safety, as fire-damp may at any moment rush out of a hole, and if set alight it would go off like gunpowder or gas from coal, killing everybody within its influence, and bringing down the tops and sides of the passages.

In some mines where it is important to have ventilation, there are four shafts, two up and two down-cast. The latter, where the coals are drawn up to the surface, are in the lowest part of the mine, and all the passages are on a gentle ascent towards the furnace, so that the air down the shafts is drawn that way. The furnace consists of a number of iron bars placed horizontally across the end of a large brick arch, and the roof and sides are built of the best fired bricks. On the iron bars nearly a ton of coals is kept constantly burning and throws out a great heat, relays of men being employed in replenishing it. At the back of the furnace is a shaft to carry off the smoke. Thus the cool air circulates all over the mine. When a large supply of air is required in any particular part of the mine, the doors are closed at the entrance to the other parts, thus directing the current where it is most wanted. This current is so strong that on opening one of these doors, care is necessary in shutting it, as it would slam with a force sufficient to knock a man down.

These and other arrangements, and the vast amount of machinery now employed, had not, however, been introduced when Mark Gilbart began life as a "trapper." The most dangerous operation is the opening of a new passage, from which foul air may suddenly escape and poison the miners inhaling it, or a stream of water may rush forth, rilling up the gallery, and drowning all within its reach. Numberless, indeed, are the dangers to which miners are exposed. Their condition is now improved, but they formerly worked eleven or twelve hours a-day, and occasionally even from thirteen to sixteen, far down in the depths of the earth, in a heavy and noxious atmosphere, in a half naked state and in unnatural positions, kneeling, stooping, lying upon their sides and backs, at any moment liable to the loss of life. The miner has not only to undergo bodily labour, but must exercise skill, patience, presence of mind, coolness, and thoughtfulness. Countless, also, are the dangers to which they are exposed. To accidents as they come down or go up the shafts by the breaking of ropes, or the giving way of machinery, from the falling in of the roof or walls, as also from accidents in blasting, from spontaneous combustion, from explosion of fire-damp, suffocation from choke-damp, and eruptions of water, and even quicksands. Sometimes floods or heavy rains find their way down unknown crevices into the pit, where the miner is working, and forming a rapid torrent, suddenly inundates the mine and sweeps all before it.

Such was the life young Mark Gilbart was apparently doomed to lead.

Chapter Three
Learning to Work

We must proceed more rapidly than heretofore with Mark Gilbart's history. He did his duty as a trapper, never falling asleep, and always opening and shutting the trap at the proper moment. The rolley boys never complained of him, and as he was invariably in good humour, and stood their chaffing, he became a favourite.

Often he had to go into the pit before daylight, and remain until ten o'clock at night with one candle to light him on his way to his trap, and another with which to return.

As he always told his mother that he was happy, and he appeared to be in tolerable health, she became reconciled to his being thus employed, though she little dreamed of what he had really to go through. When he had shorter hours of work, he employed his time at home in reading and improving himself in writing. He had also a fancy for making models. He began by making one of the parts of the pit in which he worked. Then he tried his hand at making some of the simpler machinery of the pit. His uncle acknowledged that the rolleys, corves, picks, and spades were wonderfully exact,—indeed, was so well pleased that he allowed him a lantern and a supply of candles, so that instead of sitting in the dark, he could pass his time in reading and cutting out his models, the materials for which he carried down with him. So perfect were his models that they were readily purchased by visitors to the pit. His mother, on one occasion, taking some of them into a neighbouring town, sold specimens to tradesmen, who offered to buy as many as she could bring them of the same description. At length Mark became big enough to be a "putter," or rolley boy. He could no longer read or make models down in the pit, but he got better wages, shorter hours of work, and his health improved with the exercise. Being always wide-awake, he escaped the accidents from which so many of his companions suffered, which they called "laming." The injuries they received were from various causes, but generally from falling, when the rolley passed over their arms or legs, and broken limbs were the consequence. Some had lost one or more fingers or toes, others had received gashes in their faces, or arms, or legs, but they had seldom long been laid up, and had willingly again returned to their work. The term "putter," it should be understood, includes the specific distinction of the "headsman," "half-marrow," and "foal." The "headsman," taking the part of conductor, pushes behind. The "half-marrows" drag at the sides with ropes; while a "foal" precedes the train, also dragging by a rope. Mark, however, was not very long employed in this laborious task, for the overseer, hearing of his talent, appointed him to the duty of "crane-hoister." The term explains itself. He had to hook on the "corves," and keep an account by chalking on a board the number hoisted up. In this occupation he was able to gain a pound a week. Some part of this he laid by, and with the other he enabled his little sister to attend a respectable school in the neighbourhood, where she made great progress, and showed a considerable talent for music. Mark had by this time gained the esteem not only of his companions but of the under-viewers, and was favourably known to the viewer. On several occasions when his services had been required, he had accompanied one of the under-viewers on his visits through the mines. He thus traversed the main gallery, the side walks, and the old, or abandoned works. In the latter the roof was propped up by perpendicular posts and horizontal beams. In many places the beams were so bent by the weight of the superincumbent earth, that it appeared they must before long give way. In many places they had to creep on hands and knees to pass through the old workings, which opened into others farther on.

As they made their way along, the under-viewer showed him a fault in the coal seam, and explained what it was. Coal seams generally run in a parallel position with the various other strata for a considerable distance, when, all at once, they abruptly terminate. This is marked as plainly as if a wall had been built up at the end of the seam. Thus, while on one side of the wall there is a thick seam of coal, on the other there is a mass of rock. This break or fault was caused at some remote period of the world's history by an internal convulsion. It is known, however, that the seam will again be found, either at a higher or lower level than the one first worked. To reach the seam a tunnel is driven right through the rock, when sooner or later the seam is discovered. In the present fault, a tunnel had been run through the solid rock for fifty feet in length; and they might afterwards have to follow up the seam, extending perhaps half-a-mile, or even a mile, for the whole of which length a gallery would have to be cut, from which, side workings would extend on either side. So accurately did Mark note all he saw, that on his return home he was able to draw out a plan of the mine, with which the under-viewer was so pleased, that he took it to the manager.

"This boy deserves encouragement. We must see what can be done for him!" was the remark. Shortly after this, great improvements were introduced into the mine. Fresh shafts were sunk, for affording better ventilation, and for more rapidly getting the coal to the surface. Near them, engines of great power were placed to perform the various operations required. An endless wire rope was made to run from the shafts to the extreme end of the gallery, kept revolving by a steam-engine down in the mine. The man walking ahead of the leading waggon, to which is secured a pair of iron tongs, grips hold by them of this endless rope, which thus drags on his waggons without any labour on his part, towards the shaft, up which the coals are to be carried to the surface. The chief gallery was divided by a wall down the centre, with openings at intervals of twenty yards or so, to enable persons to pass through. There were also niches on either side, where he could stand while a train was passing. On one side of the gallery the full trains ran along on rails from the workings to the shaft; on the other side the empty waggons returned to the workings to be filled. For the purpose of better ventilating the mine, an enormous fan, forty feet in diameter, formed like the paddle-wheels of a steam-ship, and kept constantly revolving by steam-power, was placed over a shaft sunk for that sole object. The suction caused by the enormous paddles drew up all the foul air and noxious vapours from the whole of the mine, and at the same time drew in from another shaft, more than a mile distant, a current of fresh air, amounting from 70,000 to 80,000 feet per minute, thus doing the work of a furnace far more effectually, and at much less cost.

Instead of the old corve or basket, an iron safety-cage had been introduced, sliding up and down on steel bars, resembling indeed a perpendicular rail-road. Wonderfully changed was the appearance of the mine itself. Mark, who had been employed above ground for some time, was astonished, on being lowered in the new safety-cage, to find himself on stepping out at the bottom in a spacious brick-arched vault, almost the size of a railway terminus, well lighted by large glass lamps suspended from the roof. The machinery, both steam and hydraulic, looked in the most perfect order; the steel parts of the engine shining like burnished silver. Trains of laden waggons were every now and then arriving. First of all was heard a distant rumbling, with the "whirr" of the iron rope far back in the darkness. The rumbling sound grew louder, and at last the train came in sight. A stalwart miner, with his lamp dimly twinkling slung at his waist, striding along holding in his left hand the iron tongs before mentioned, and having behind him a long train of waggons, gradually came into the light. On he went to the foot of the shaft. Here a strong iron cage appeared, having three floors, one above the other. In front of this was a stage, on to which the leading waggon was run. It was then lifted by hydraulic power, until a second stage appeared below it. On this another waggon was run, that again rose, until a third stage was level with the tramway—the three stages being now level with the three floors of the cage. At the same time three hydraulic rams or arms ran out from the side of the shaft and pushed the waggons into the cage, which immediately began ascending. It should have been said that three empty waggons had come down in the cage, and had in the first instance been withdrawn and placed on the return tramway. These were at once coupled together by men stationed there for the purpose, who had now to wait for the return of the cage with more empty waggons to be again filled with three others from the full train. The cage on reaching the summit of the shaft was unloaded much in the same fashion by hydraulic power. This operation was carried on with wonderful rapidity, so that the outputs, or amount of coal raised, averaged from 800 to 900 tons per day.

More than a mile away from this main shaft was the engine-room which worked the endless rope. On a platform some distance above the ground sat the engineer, surrounded by a multitude of signals. In spite of the tremendous noise which prevented one person hearing what another said, the engineer attended to all his signals with the greatest accuracy, his complicated machinery in beautiful order, and appearing perfectly at his ease. Some idea may be formed of the vast amount of labour employed in this mine when it is understood that the working-faces, with gate-roads, main roads, air-ways, returns, engine-plains, self-acting and engine inclines, extended upwards of eleven miles, and with the addition of the old working roads, including those which were bricked up, the whole measured the enormous amount of twenty-two miles. All these passages were kept far better ventilated by the fan than they were by the furnace hitherto in use, while the pure air brought down, greatly contributed to the health of the miners.

Mark had risen step by step. He was now able to take a house for his mother and Mary, although old Hayes and his wife were very unwilling to part with them. Mary had greatly improved in her music, of which she was passionately fond, but she had no piano on which to play at home.

Mark, who had a holiday, hearing that an auction was to take place at the neighbouring town, at which a pianoforte was for sale, set off to attend it. There was some competition, but he had 20 pounds in his pocket, saved from his earnings, and it was finally knocked down to him at that price. With proud satisfaction he at once hired a spring cart, and set off with it for his home, where he had it placed while Mary was out with their mother. Her delight at seeing it equalled the pleasure with which he bestowed his gift. The fact was inserted in one of the local papers by the auctioneer who sold it, that the piano was purchased by the first 20 pounds saved out of the earnings of a collier boy, as a present to his sister.

Unhappily, such instances are rare, for although many collier boys gained high wages, the money was too generally lavishly spent, without thought for the future.

Of late years a considerable improvement has taken place among many mining populations, but even in former years it was possible for talent to force its way upwards. Who has not heard of George Stephenson, who began life trapper in a mine at six years of age, and rose to be a great engineer, father of Robert Stephenson, M.P., and engineer-in-chief of the North-Western Railway; of Dr Hutton, who was originally a hewer of coal in Old Long Benton Colliery; of Thomas Bewick, the celebrated wood-engraver; of Professor Hann, the mathematician, and of many others whose names are less known to fame, who have obtained respectable positions in society.

Old Hayes had lately moved to another pit some distance from the one in which he had hitherto laboured, being tempted by higher wages, and Mark shortly afterwards was offered a situation as under-viewer in the same pit. It was worked on the old plan, but improvements were being carried out.

Old Simon with four other men were coming along the main gallery, being the last of the miners who were leaving the pit for the night. The rest had already gained the foot of the shaft, when a rushing, roaring sound was heard followed by a tremendous blast of wind, which, almost took them off their feet. The cage was at the bottom of the shaft. They sprang into it, more than double the number it usually contained clinging on. Before they could give the signal to be drawn up, they saw a torrent of water surging on several feet in depth, rapidly filling the whole lower part of the mine.

They were soon out of danger, but what had become of old Simon and his companions? Mark had come to the pit's mouth intending to descend and make his usual survey of the mine to see that all was right. He soon heard on inquiry of the supposed fate of old Simon and the rest. No one doubted that he had been overwhelmed by the raging waters, but that such was the case Mark was not thoroughly satisfied.

"They may have escaped in one of the side workings, and if so they are still alive, although it may be a difficult matter to get them out," he remarked.

He at once ordered the cage to be lowered, and with two men who volunteered to accompany him, descended in it. On getting near the bottom he discovered that although the water had filled the main tunnel to the roof, there was still a passage running away to the left on a higher level which was perfectly dry. They proceeded along it although his companions considered that a search in that direction was useless.

"If the poor fellows were last seen in the main gallery, it seems impossible that they should have got up here," they remarked. They, however, went on and on, but no signs of human beings could be discovered. They were returning, and were once more approaching the shaft, when a dull sound was heard, as if some one was striking on a wall in the far distance.

Mark placed his ear against the side from which the sound seemed to come, and he distinctly heard several blows given. The others did the same.

"You are right, Gilbart, that comes from the side working nearest to us. The men must be there," exclaimed one of his companions.

"We will reply to them," said Mark, and taking a pick he struck several heavy blows against the side of the gallery. They were replied to by the same number.

"How is it that they can be there and not be drowned?" asked one of the men.

"The water is prevented from rushing in by the pent-up air in the working," he answered. "How long it will be kept back I cannot say, but no time must be lost in hewing a way through to them. Come, lads, with God's help, we will save them," said Mark. "Keep picking away until I return," and he hastened to the shaft.

Having an exact plan of the mine, he was able to determine at once the working in which old Simon and his companions were imprisoned. The distance, however, to the spot where he was convinced they must be was fearfully great, between eighty and ninety yards. It would take days to bore through. Would those they desired to save be able to exist so long? The attempt must be made.

Volunteers were quickly obtained, and descending with a dozen skilful hewers, he commenced operations at the very spot where the sound of the blows had reached his ears. In a short time a gang of putters with a supply of rolleys came down to carry away the coal and earth and rock as it was hewn out, but five men could only labour at a time. They worked, therefore, in relays. Day and night they laboured on without cessation, except occasionally stopping to ascertain that their friends within were alive, when they were encouraged to proceed by invariably hearing the knocking which had at first attracted Mark's attention. He directed the course they were to pursue, never once ascending to the pit's mouth, but taking his food near the working, and sleeping in a blanket on the hard rock. Day after day and night after night they worked on. The knocking from within sounded louder. On the seventh day their leader, an old friend of Simon's, struck his pick into the rock before him, making a deep hole, through which there suddenly rushed out a stream of noxious gas, and he fell overcome. His comrades, seizing him by the arms, dragged him out, narrowly escaping themselves. Reaching the fresher air, he soon recovered, and undaunted exclaimed, "Let me go at it again, lads!" and leading the way, once more the bold miners recommenced operations. Still another day they worked on, and the partition which divided them from their friends was growing thinner and thinner. A second escape of gas once more compelled them to retreat, but as soon as it had dispersed, with the courage of heroes they again went at it. At length, on the tenth day since the water had rushed into the mine, but a thin wall remained between them and the imprisoned ones. They had now come to the most dangerous part of their undertaking, the moment they had broken away the wall, the compressed air would rush through the aperture, with a force far greater than the fiercest hurricane, and the water surging up might drown those within. Still, they knew they must risk it.

"Now, lads, we'll do it," cried their old leader, and lifting his pick he struck a blow against the rock. As he withdrew it, the air rushing through extinguished the lights, and they were left to work in darkness. Notwithstanding this, in spite of the wind in his face, the old man worked on with thundering blows. Every moment he brought down masses of rock until he was convinced that he had made a hole large enough to creep through.

"Where are you, lads?" he shouted. "Come on, come on!"

Some faint voices replied, he and four others, clambering through the aperture, each lifted a man in his arms. They could hear the water rushing in close to them, but they hesitated not. Dragging out their friends, they staggered along the gallery they had just formed. They were met by Mark and a party of men carrying lanterns, and battling against the fierce blast which rushed through the passage. They were thus soon relieved of their burdens. Quickly reaching the main gallery, the doctor took the rescued men in hand, having a plentiful supply of food, medicine, and attendants ready. Though weak and almost exhausted, the five men in a few hours were sufficiently recovered to be conveyed up the shaft, where they were received by their relatives and friends, who long before had given up all hopes of ever again seeing them.

It may be asked how were these men able to live so long during their imprisonment! Fortunately they had with them a small store of provisions, and knowing that it might be many days before they could be rescued, they at once put themselves on the very smallest allowance that would support life, at the same time the air, which as we have seen was so compressed by the force of the water, was capable of sustaining respiration for a much longer period than when of its ordinary density.

There is a very great amount of vitality in the human frame, and where the wear and tear of active labour does not exist, man can live for a long period almost without solid food, especially if there be a plentiful supply of fresh water at hand.

Chapter Four
The Mines of Europe

Mark Gilbart had never thrown a moment away. By study, perseverance, and strict integrity, and the exercise of the intelligence with which he was endowed, he had risen step by step to a far higher social position than he had before enjoyed. Though still young, he had become a mining engineer, and was greatly respected by all who knew him. He had the happiness of placing his mother and sister in a house of their own, without the necessity of labouring for their support.

He was one day drawing plans in his study, when he received a note from a Mr Harvey, a gentleman of property, the owner of several mines, requesting him to call.

Mr Harvey received him cordially. "I am about to ask you, Mr Gilbart, to accompany my son Frank on a tour of considerable extent, to visit some of the more important mines in Europe, and, if there is time, in other parts of the world, and he is anxious to have a practical man who will enable him to comprehend the different matters connected with them more clearly than he would be able to do by himself. I need not say that I am fully aware of the value of your time, and I therefore offer you such compensation as I hope you will consider sufficient."

Mark gladly agreed to the proposal. Such a tour was above all things such as he desired, and which, indeed, he had himself contemplated taking at his own cost. Frank Harvey was an active, intelligent, young man, exactly the sort of companion Mark would have chosen. Having concluded all their arrangements, they lost no time in setting out.

Having visited the English, Scotch, and Welsh coal districts, numbering in all about fifteen, they bent their steps—after seeing the iron and lead mines in the south of Scotland, and the north and centre of England—towards Cornwall, to explore its tin and copper mines; after which they intended to cross the Channel to visit the more remarkable ones of Europe.

Their first halting-place was at Redruth, near which is the lofty hill called Cairn Brea, whence they obtained a view over an extensive mining district. The country around, covered in many places with enormous blocks of granite, looked barren and uninviting in the extreme, and no one would have supposed that any portion of the soil in sight was the richest in the whole of our island. Within a few miles of the spot where they stood were, however, numerous copper and tin mines, many of which had yielded a large profit to their owners. Among them was Dolcoath, one of the oldest copper mines in Cornwall, 300 fathoms in depth. Another, Eastpool, a tin and copper mine, from which ores to the amount of 130,000 pounds have been won, after an original outlay of only 640 pounds. From the former mine native silver, cobalt, and bismuth have also been obtained. The mineral deposits of Cornwall, it should be known, are found in granite and grey slate. Those of Derbyshire and the north of England—lead and iron—in the carboniferous system.

The travellers visited these and several other mines, among them the Consolidated Copper Mines, situated in the parish of Gwennap, about three miles from Redruth. They extend along the brow of a range of steep hills, into which numerous shafts are sunk. The length of the whole of these shafts together, it is calculated, is more than twelve miles in perpendicular depth, and if to these are added the horizontal galleries, which perforate the hill in all directions, the extent of subterranean excavation is upwards of sixty miles.

Eight steam-engines of the largest size, and thirty of smaller dimensions, are employed for drainage and other purposes, their ordinary working power being equal to 4000 horses, but when their full power is put on they almost equal that of 8000. To carry off the water from these mines, a tunnel, with numerous ramifications has been formed, measuring nearly thirty miles in length. One branch of this tunnel is upwards of five miles long, carried underground 400 feet beneath the surface, finding its outlet into the sea near Falmouth.

A few years ago the number of tin mines worked in Cornwall amounted to 139, and to 26 in Devonshire; and about 20,000 persons were employed in them.

Although the wages of the miners are much inferior to those of the pitmen in the northern coal-fields, yet they have advantages over their brethren, being exempted from many of the evils to which the northern miners are subjected. They have no fear of the fatal fire-damp or sudden explosions. Intellectually they are also superior, as they are mostly engaged in work requiring the exercise of mind. Their wages arise from contract, and depend greatly upon their skill and energy. They mostly have gardens, which they cultivate, and when near the coast they engage in the fisheries, thus increasing their incomes and varying their mode of life.

After leaving Redruth the travellers proceeded over the wildest and most desolate of moorlands, with blocks of stone scattered about, towards the wonderful Botallack Mine, on the Cornish coast. No mine in the world is so singularly placed. Descending to the shore below, on looking upwards, the view appeared fearfully grand. In one part was a powerful steam-engine, which had to be lowered almost 200 feet down the cliffs. Here tall chimneys, pouring out dense volumes of smoke, were seen perched on the ledges of a tremendous precipice. Here and there also were the huts of the miners, disputing the ground with the wild sea-birds, while ladders of great length scaled the rocks in all directions, enabling them to ascend and descend to their work. In some parts were paths up which sure-footed mules, with riders on their backs, were trotting briskly along, where few people unaccustomed to dizzy heights would have wished to venture even on foot.

As they had determined to visit the mine, they had to ascend to the top of the cliff and then once more to descend among the rugged rocks to a ledge about midway between the summit and the ocean, where a small building, occupied by the mining agent, marked the entrance. Hearing who they were, the agent at once undertook to guide them, and produced a couple of woollen mining dresses and two large felt hats.

CORNWALL MINES BELOW THE SEA.

Each person having fastened four or five candles to his button hole, while he carried another in his hand, they began to descend through a trap-like entrance, by a series of ladders, which although strong enough in reality had a very rickety feeling. On reaching the foot of one ladder, they were conducted to the top of another, on to which they had to step, and thus descending ladder after ladder and passing ledge after ledge, they at length reached the bottom of the pit, where the end of a pump was seen drawing up the water from all parts of the mine.

They then commenced their progress along one of the numberless galleries, which was so narrow that two persons could scarcely pass each other. Now having to step over rough stones and often close to the edges of fearful pits, now to bend low under masses of overhanging rock, and sometimes to find themselves crossing unknown abysses by shaky bridges of planks, while the damp air felt hot and sickly, making the candles burn dimly. Here miners were at work with pickaxes getting out the ore. Having thus gone over, through, and under all impediments, they were informed that they were 120 feet below the level of the sea vertically, and horizontally 480 feet below low-water mark. Boats might even then be passing over their heads. Human beings were working still lower down. On the roof, the strips of pure copper could be distinguished among the crevices of the rocks through which the salt water was seen percolating in an unpleasant abundance. In their eagerness to obtain the rich ore, the miners had worked upwards until they had got within five or six feet of the bottom of the ocean. There the metal was still clearly visible, but even the most hardy miners would scarcely have ventured on an attempt to win another grain from the rock overhead, lest the water should rush in and overwhelm them, and inundate the mine.

Passing into a gallery where no one was at work, the travellers listened in perfect silence, and could hear the low murmur of the ocean rolling above their heads.

"Oh, that is nothing now," said their guide. "When a storm is raging, I have heard the sound of the pebbles, which some large wave has carried outwards, bounding and rolling over the rocky bottom. On standing beneath the base of the cliff, where not more than nine feet of rock intervened between the sea and my head, the heavy roll of the large boulders, the ceaseless grinding of the pebbles, the fierce thunder of the billows with the crackling and boiling as they rebounded, produced an uproar such as those who heard it can never forget."

For many years a blind man worked in the Botallack Mine, and supported a large family by his labour. So complete was his recollection of every turning and winding, that he became a guide to his fellow-labourers, when by any accident their lights were extinguished. He being afterwards cruelly discharged, engaged himself as an attendant to some bricklayers. While thus employed, with a hod of mortar on his back, he fell from a platform and was killed.

There are several other mines similarly situated to that of the Botallack on the coast of Cornwall, where the works are carried far under the ocean. Among them are the Wheal Edward, the Levant, the Wheal Cock, and the Little Bounds. In the two latter, the miners have actually followed the ore upwards until the sea itself has been reached, but the openings formed were so small that they were able to exclude the water, by plugging them with wood and cement.

On returning from the mine, the travellers, having doffed their miners' dresses, inspected the outward machinery employed in crushing the ore on the landing-place in the side of the cliff, and drawing it up the precipitous tram, which leads to the summit, where it is stamped and prepared for exportation. It is mostly carried to Swansea, which, in consequence of the abundance of fuel in the neighbourhood, owing to its nearness to the sea, to its canals and railroads, has, in the course of half-a-century, from a mere fishing village become a town containing fully 40,000 inhabitants.

The Cornish mines are not the only ones which run under the sea. On the Irish and some parts of the English coasts there are several coal mines which are worked beneath the ocean bed to a great distance.

Another remarkable mine, that of Huelwherry, existed for many years on the Cornish coast. A rocky spot at about 120 fathoms from the beach was left dry at low-water, on which small veins of tin ore were discovered crossing each other in every direction. Although the surface was covered for about ten months in the year, and had at spring-tides nineteen feet of water over it, while a heavy surf often broke on the shore, a poor miner, named Thomas Curtis, about a century ago determined to attempt winning the ore. The work could only be carried on during the short time the rock appeared above water. Three summers were spent in sinking the pump-shaft, which had every tide to be emptied of water.

A frame of boards, raised to a sufficient height above the spring-tides, and rendered water-tight by pitch and oakum, was placed above the mouth of the shaft. Its sides were supported by stout props in an inclined direction. At the top of this wooden construction, which was twenty feet in height, a platform of boards was secured, on which a windlass was placed. The water was now pumped out of the mine and the machinery set to work; but the sea penetrated through the fissures of the rock, and greatly added to the labour of the workmen, while during the winter months, on account of the swell, it was impossible to convey the tin ore to the beach. Notwithstanding all these difficulties, the persevering projector was rewarded by obtaining many thousand pounds worth of tin. At length, during a gale, an American vessel broke from her moorings, and demolished the machinery by striking against the stage, when the water rushing in filled the mine. An attempt has been made of late years to again work the mine with improved machinery, but the venture not proving profitable it has been abandoned.

The travellers also visited the curious Carclaze tin mine near the town of Saint Austell. It is a prodigious hollow or basin, nearly thirty fathoms in depth and a mile in circumference, and has the appearance of a natural crater rather than a hollow made by human hands.

The sides are almost perpendicular, and a few footpaths alone lead down amid the rocks to the bottom. In every direction are seen the hollows made by the miners of ancient days, the white colour of the granite veined with the darker metalliferous streaks, and the curious shape of the rocks formed by the streams flowing down its sides, give it a remarkably picturesque appearance.

The machinery used for crushing the rock is set in motion by these streams. On every side the men, women, and children employed on the works are seen moving about in all directions, like a busy colony of ants. The ore is obtained without much difficulty.

A tunnel has been formed at the bottom of the mine through which the waters flow after they have performed their task, which also carries away the crushed granite, while the heavier metalliferous substances are precipitated into the troughs. Neither engine-house nor chimneys such as are seen in other mines are visible, while every detail of the work is exposed to view—indeed, the huge basin has the appearance of a mine completely turned inside out.

There are two methods of smelting tin. By the most common, the ore, mixed with culm, is subjected to heat on the hearth of a reverberating furnace, when ordinary coal is employed. By the other method, the ore is fused in a blast furnace, when wood fuel or charcoal is used. The tin when smelted runs off from the furnace into an open receiver, from which it flows into a large vessel, where it is allowed to settle. After the scoriae have been skimmed off, the upper and purer portion of the mass is refined, and the lower part re-melted.

Chapter Five
The Metals found in Mines

The chief object of the travellers was to inquire into the mode in which mines in different countries are worked, the causes of accidents, and the best method of preventing them. Their knowledge was superior to that which most of our readers are likely to possess, and it will be necessary, in order to understand their proceedings, to glance at the mining districts of the world, and to describe some of the principal mines among them.

No country possesses, within the same area, so large an amount of varied mineral wealth as Great Britain. Besides the seventeen coal districts of Great Britain, we find in Scotland numerous lead mines in the clay slate mountains on the borders of Lanarkshire and Dumfriesshire. In the north of England, with Alston Moor as the centre, along the borders of Northumberland, Cumberland, Westmoreland, and Durham, are extensive veins of lead. Cumberland, the north of Wales, and the Isle of Anglesey produce copper ore, as also mines of lead and magnesia, with many other metals,—zinc, arsenic, cobalt, and bismuth. Iron in large quantities is found in South Wales, South Staffordshire, and in the Scottish coal-fields, where the ironstone appears in abundance alternating with layers of coal and other strata, and is generally won from the same pit as that from which the coal is extracted.

Besides coal, Ireland contains mines of copper and lead, found in the slate and limestone ranges, contiguous to the sea coast. Crossing from thence to Spain, we arrive in a country rich in mines, though, owing to its distracted state, for many years greatly neglected. Here lead is found in large quantities in the mountain chains.

Quicksilver is abundant from extensive veins of cinnabar in the province of Mancha. In Galicia tin has been produced from very early times. Iron ore is very abundant, and silver mines, for many centuries abandoned, are now again being reworked. Gold was at one period discovered in large quantities, but is supposed to be almost exhausted.

The most important coal-field of France is round Étienne, near Lyons. Mining operations are also carried on in Brittany and the Vosges. Although possessing less mineral wealth than England, the French were far in advance of us in regard to the management of their mines. Germany possessed the chief school for scientific mining. Its principal metalliferous sites are the Hartz Mountains, on the borders of Hanover and Prussia, and the Erzgebirge or Ore Mountains, which separate Saxony from Bohemia. They yield silver, copper, lead, iron, tin, and cobalt.

The most prolific sites of the precious metals in Europe are possessed by Austria. The Styrian Alps furnish a vast amount of iron. The province of Carniola supplies quicksilver. Hungary and Transylvania, copper, lead, antimony, and iron. The most extensive works are found in the neighbourhood of the town of Cremnitz and Schemnitz. The veins in this region obtain the enormous dimensions of from 20 to 200 feet in width. The extensive forests of oak, pine, and beech which clothe the hills supply fuel for the numerous smelting works, while water, carefully collected into reservoirs, moves the required machinery. The whole of the drainage of the mines is collected in a receptacle 600 feet below the surface, from whence it is conveyed under a lofty mountain ridge by a magnificent gallery twelve miles in length.

Norway and Sweden possess extensive mines of iron and copper, as also silver. The latter country furnishes the best iron in the world, and it is much used in England for the manufacture of steel.

Passing eastward to Russia, we find the rich mines of the Ural Mountains, which divide Europe from Asia, and then on to the Altai chain on the southern frontier of Siberia, we meet with rich mines of gold and silver, and other valuable metals. On the European side of the Ural there is a deposit of copper sand-ore, extending over a district of 480 miles in length, by 280 in breadth. The mineral wealth of Asiatic Russia is far greater. It consists of copper ores; iron cropping out at the surface, gold and platinum. The Altai Mountains especially produce silver, and some gold, with lead and copper ores. The silver mines of this region were worked at a very early period, as is proved by the discovery of an excavation a thousand feet in length, from which a stone sphinx was dug up, corroborating a statement of Herodotus that the Scythians possessed mines of gold and silver, which, according to his account, were guarded by monsters and griffins. Baron Humboldt supposes that he referred to the bones of elephants, and other gigantic animals, discovered at the present day in the steppes between the Ural and Altai chains.

Crossing the Atlantic to America, we find vast quantities of the precious metals in the mountains of the Brazils and along the whole range of the Andes. In the province of Minasgeraes, gold is obtained from subterranean excavations, as also by washing the surface soil, when diamonds are also found. Auriferous deposits exist in the deep valleys among the mountains of Chili, and in Peru and Bolivia are immense veins of silver ore. High up on the Andes are the mines of Pasco and Potosi; while in the same region, quicksilver, copper, lead, tin, and other metals have been discovered. The copper mines being nearest the sea, are generally worked, the ore being sent to Swansea.

The lofty plateau of Mexico in North America has, from the first, been celebrated for its rich silver mines, of which there have altogether existed no less than three thousand, but the larger number of these have long been unworked. The gold mines of California and of Australia are too well known to require mention; but we must not forget the rock oil, concealed for ages in the North American continent. Both the United States and Canada now yield an abundant supply.

The number of metals discovered beneath the surface, including the metallic basis of the earth and alkalies, amounts to forty-two, but metals, commonly so-called, number only twenty-nine. These are platinum, gold, tungsten, mercury, lead, palladium, silver, bismuth, uranium, vanadium, copper, cadmium, cobalt, arsenic, nickel, iron, molybdenum, tin, zinc, antimony, tellurium, manganese, tatiaum, chromium, columbium, rhodium, iridium, osmium, cerium. Many of these, however, are so rare, that as yet they are of no practical use. Gold has been known from the earliest ages, and is found in scales, threads, grains, and rolled masses, or nuggets, which latter have been discovered in California and Australia weighing from twenty to thirty pounds, but the largest of all met with was in Asia, on the southern side of the Urals.

Large quantities of gold were discovered on a marshy plain which had been thoroughly turned over, when it was resolved to take down the buildings in which the gold was washed, and under the very corner of one of them a lump was found, weighing no less than ninety-six and a-half pounds troy, and valued at 4000 pounds. Gold has been found in Scotland, and in the county of Wicklow, Ireland, where about 10,000 pounds worth was picked up in the bed of a river by the inhabitants, before the Government became aware of its existence. Gold is so malleable that a single grain can be beaten out to form a gold leaf covering a surface of fifty-six square inches, and it is so ductile that the same quantity may be drawn into a wire 500 feet in length. Silver is found embedded in various rocks, where it occurs in veins, assuming arborescent or thread-like forms, and occasionally appearing in large masses. The largest mass found in Europe was brought from Kongsberg, in Norway, weighing upwards of 560 pounds, but another, won from the mines of Peru, was said to weigh 800 pounds. The celebrated mines of Potosi, 10,000 feet above the sea, were discovered in 1545 by an Indian who, when chasing a deer, laid hold of a shrub to assist in his ascent; it came up by the roots, to which he found attached a quantity of glittering particles, which he at once knew to be silver. Veins of silver have been discovered in England and Scotland, but generally mixed with lead.

Iron, the most useful of all metals, is found in large quantities in England, in many parts of Europe, and the United States. At one time Sussex was full of iron mines, the furnaces being fed with charcoal, until so extensive was the destruction of the woods and forests that the Government interfered, and placed restrictions on the consumption of the timber.

On the discovery of the present method of smelting with pit coal, the works, which at one time numbered 140 in Sussex alone, were abandoned, and hop-fields now cover the ground where furnaces once blazed.

Copper ranks next to iron in utility. In Cornwall there are upwards of 100 copper mines. It derives its name from the island of Cyprus, where it was first obtained by the Greeks. It is employed pure for numerous purposes, and is also mixed with other metals to form bell metal, speculum metal, for optical purposes, and German silver.

Lead occurs in veins most plentifully in mountain limestone districts, and usually contains some portion of silver. There are lead mines in various parts of England, as well as in Spain, Saxony, and in Bohemia, and some very rich lead mines have of late years been worked in the United States.

Tin is found in Cornwall in larger quantities than in any other part of the world. It is generally discovered in the alluvial soil of low grounds, where it is known as stream tin, because it has been washed by the agency of water from the rocks in which it was originally embedded mixed with sand and gravel. Tin is also found in the island of Banca, in the Indian Archipelago, in Bohemia and Saxony, Chili and Mexico.

Mercury is a rare metal. The richest mines are at Almaden, a small town of La Mancha, in Spain. It is also found in Austria, China, and Peru, and a few other places. It is sometimes found in globules, but it is generally procured from one of its ores, cinnabar, a sulphate of the metal, of a red colour, and indeed identical with the richly prepared paint vermilion. A thousand workmen are employed in the Spanish mines, above or under ground. It freezes at an exceedingly low temperature, and was found solid during midwinter by the traveller Pallas. Of the other metals, some used as medicines, or pigments, or to form alloys, we have not time now to speak.

Chapter Six
Salt and Quicksilver Mines

The object of the travellers was not only to inspect coal mines, but to view the wonders of the subterranean world. It is impossible to do more than give a very brief account of the places they visited. They had found their way to the Carpathian Mountains, in order to visit the salt mines of Wieliczka, a small town to the south of Cracow. The valley in which the mine is situated is fruitful and picturesque. Descending by a staircase of thirty feet or so, through a bed of clay, they arrived at the commencement of the level galleries, which branch off in all directions. Overhead was a ceiling of solid salt, under foot a floor of salt, and on either side grey walls of salt, sparkling here and there with minute crystals. The guide led them on through a bewildering maze of galleries. Now they entered a grand hall, now descended by staircases to another series of vaulted chambers. On every side was solid salt, except where stout piers of hewn timber had been built up to support the roof, or wooden bridges had been thrown over some vast chasm. As they descended, the air became dry and agreeable, and the saline walls more pure and brilliant. One hall, 108 feet in length, resembled a Grecian theatre, the places where the blocks had been taken out in regular layers representing seats for the spectators. Here and there were gangs of workmen, some labouring at the solid floor, others trundling wheel-barrows full of cubes of salt.

Soon after entering, they reached the chapel of Saint Anthony, excavated in the times of the Byzantines, supported by columns, with altar, crucifix, and life-size statues of saints. They appeared, from being coated with smoke, to be of black marble, but Mark, putting his tongue to the nose of one of the saints, discovered it to be of salt. Many of the saints, however, were disappearing before the damp, which enters in that higher region from the upper world. The heads of some, and the limbs of others had already fallen.

The guide had come provided with some Bengal lights, one of which he kindled on the altar, bringing into light this strange temple; then, as the flame burnt out, the whole vanished as if by magic. Passing across a wooden bridge, resting on piers of salt, they entered a vast irregular vault in which were two obelisks of salt, to commemorate the visit of Francis the First and his empress. As they reached the floor, a boy ran along the bridge above with a burning Bengal light, which threw flashes of blue lustre on the obelisks, the scarred walls, the vast arches, the entrance to the deeper halls, and the lofty roof, fretted with the picks of the workmen. Another hall was entered, with cavernous tunnels at the farther end, passing through one of which, they embarked upon a lake in a heavy, square boat, and entered a gloomy passage, over the entrance to which was inscribed, in salt letters, "Good luck to you." Midway in the tunnel the halls at either end were suddenly illuminated, and a crash, as of a hundred cannon bellowing through the vaults, shook the air in such a way that the boat had not ceased to tremble when they landed in the farther hall. The noise was produced by a single gun.

A tablet, on which was inscribed, "A hearty welcome," greeted them on landing. At a depth of 450 feet their journey ceased, although they were but half-way to the bottom. About 1500 men are employed in the mines, who labour only six hours at a time, and live in the upper world. The blocks are first marked out on the surface by a series of grooves. One side is then deepened to the required thickness, and wedges being inserted under the block it is soon split off. This salt bed occupies a space of 9000 feet in length and 4000 in width, and consists of five successive stages, separated from each other by intervening strata of from 100 to 150 feet in thickness, and reaching to the depth of 1500 feet.

More than ten years ago a serious accident, which threatened the destruction of the mine, occurred. While boring, to obtain some potash salts, through an aquiferous stratum, a spring was tapped, which poured an immense quantity of water into the lower galleries. The inhabitants feared not only the ruin of the mine, but the falling in of their houses from the melting of the salt pillars; but fortunately the inundation was confined to the lower galleries, and a powerful steam-engine being set to work, the water was again pumped out, and the spring blocked up. However, so vast are the excavations that it would have taken many years to fill them.

Contrasting with the bright glitter of the salt mines of Wieliczka are the gloomy slate quarries of Saint Peter's Mount, near Maestricht, in the Netherlands, the most extensive in the world. For centuries they have been worked, both for building and manuring, and probably benefiting the agriculturist more than the architect. In spring and summer the labourers occupy themselves in their fields above ground, and not until winter approaches do they begin to burrow in the entrails of the earth.

The two travellers followed a trusty guide through those endless passages, which constantly crossed each other, either to the left hand or to the right. Darkness to be felt, silence profound, reigned everywhere, even the human voice seemed to die away without awakening an echo—the only sound to be heard being an occasional dropping of water from the roof into a small pool below.

Suddenly the guide extinguished his torch, when, bold as they were, and well accustomed to subterranean regions, a sensation of awe crept over them. Their first impulse was to feel for the wall, for in vain their eyes sought a ray of light, as in vain, also, their ears listened for the slightest sound.

Neither spoke for some minutes, and they experienced a sensation of relief when the guide relit his torch. Numbers of hapless beings have been lost in these trackless galleries, and here and there are inscriptions on the walls, notifying that a corpse was found on the ground below. One poor workman lost his way, and roamed about until his torch died out of his burnt fingers. The lamp of another was overturned, and he in vain endeavoured to find his way out of some remote gallery.

A French geologist while exploring the quarry discovered a corpse shrivelled to a mummy, the hat lying close to his head, a rosary in his hand. It was conjectured to be the body of a workman who had died more than half-a-century before, the dry air and the absence of insects explaining the preservation of the corpse. Two centuries ago four Franciscan monks resolved to construct a chapel in honour of their tutelar saint. In order to be able to retrace their steps, they took with them a large ball of twine, leaving one end secured to a spot where people were constantly passing. Their twine unwound, they at length reached a vast hall, probably not visited for many ages. Near the entrance one of them drew a sketch of the convent, and wrote beneath it the date of their discovery. When about to return, what was their horror to find that their twine had snapped. They must have searched for it in vain, for never more did they return.

At last the prior, alarmed at their absence, sent parties to explore the excavations, but so vast were they even then, that seven days elapsed before the corpses of the hapless friars were found, their faces downwards, and their hands folded as if in prayer.

During the siege of Maestricht by the French Republic, a party of the besiegers occupied the quarries. The Austrians who garrisoned Fort Pierre at the back of the mountain, formed a plan to drive them out, and tunnelling made their way towards their enemies. Although they marched silently along, their torches betrayed them, and the besiegers pouring in a volley of musketry killed a large number, made prisoners of some, and drove the rest into the depths of the cavern.

On the banks of the Nile are several prodigious stone quarries, from which the cities of ancient Egypt were built. Perhaps the largest is that of Haggar Silsibis. Here passages, broad as streets, with walls fifty or sixty feet high, now stretching straight forward, now curved, extend from the east bank of the river into the heart of the mountain, where halls have been hollowed out large enough to contain the Roman Colosseum, the rough hewn irregular roof resting upon immense square or many-sided pillars, some of which are eighty to a hundred feet in circumference. Here numerous blocks, already completely separated from the rock, appear ready to be transported; the labours of the quarry-men having suddenly been arrested by the invasion of the Conqueror, who overthrew the priests of Isis.

One of the most curious quarries of ancient days is found near Syracuse. The greater portion is a hundred feet below the level of the earth, and of vast extent, the whole hewn out of rock as hard as marble, the blocks thus obtained being employed in building Syracuse. It is converted by the monks, who have a convent above it, into a garden—a romantic and beautiful spot, as no wind can touch it. It is filled with a variety of vines and shrubs and fruit trees, among which oranges, citrons, pomegranates, and figs grow luxuriantly, and obtain an unusual size. Sicily produces sulphur in large quantities—the chief sulphur pits being near Girgenti. Most of the inhabitants are employed in them, to the neglect of the rich soil of their island; they labour away in the most primitive manner, pickaxe and spade being the only implements employed.

When a promising vein is struck, the miners set to work, and filling their baskets with the sulphur, carry it out and throw it into large heaps of a conical shape. These mounds are covered over with moist clay, some openings being left for the escape of smoke; the bottom is then ignited, and the matted sulphur flows out through grooves into pans, where it congeals in solid masses. The passages to the mines are so narrow, that persons can with difficulty pass each other; they then expand into high vaults, the roofs of which are ornamented with beautiful crystals of celestine and gypsum. On account of the excessive heat, the workmen labour in a nearly nude state, their dark brown skins sprinkled with light yellow sulphur dust, making hem look savage and strange in the extreme. Towards the end of the ıst century, the sulphur mine of Sommatin caught fire, the conflagration ausing the complete abandonment of the pit. For two years it raged, until ıe mountain, suddenly bursting asunder, a stream of molten sulphur ıshed forth, and precipitated itself into the neighbouring river. The mass f sulphur, amounting to upwards of 40,000 tons, was thus obtained by the owners of the former pit, who had believed themselves ruined.

There are sulphur mines in different parts of the world, the largest of which are in Japan, but too remote to be worked with advantage. Gypsum, or sulphate of lime, better known as Plaster of Paris, is found in prodigious quantities at Montmartre, close to that city; but as it can readily be worked without having recourse to subterranean excavation, it need not be mentioned further.

When gypsum assumes an opaque, consistent, and semi-transparent form, it is known as Alabaster. The largest quarries are near Volterra, in Italy. Here the whole population have been employed for centuries, either in cutting it out of the mine, or in converting it into elegant forms of great variety, which are sent to all parts of the world.

Great Britain possesses inexhaustible alabaster mines in the neighbourhood of Derby. Some is worked on the spot, but the finest blocks are sent to the studios of sculptors.

Quicksilver, or mercury, is among the rarest of metals. The only two important mines in Europe are at Almaden, in Spain, and Idria, in Carniola. The former, situated on the Sierra Morena, was for many years farmed off to the Fuggers of Augsburg, but are now worked either by government or private companies. This was one of the most interesting spots visited by the two travellers.

Entering a spacious tunnel, completely walled with solid masonry, they advanced into the very bosom of the mountain. Here galleries branch out in various directions, hewn in the slate forming the matrix of the vein. One of them leads to a vast circular hall, called the Boveda de Santa Clara. At one time a horse gin was employed in this hall for raising the ore, but at present this work is performed through a shaft descending to the lowest level of the mine. Convenient steps lead down from another gallery to the first working level, and thence the descent is by short ladders to deeper storeys. The galleries are of a sufficient height to allow a person to work upright. The upper ones are dry, but the lower are humid and damp, although the water is easily raised by hand-pumps from storey to storey into a large receiver, which is emptied by a steam-engine. So extremely rich are the veins, that although worked for many centuries, the mine has scarcely yet reached a depth of 1140 feet. The present quantity raised annually amounts to eighty-thousand hundredweight of pure mercury. The ore known as cinnabar is of a dark-red colour, and gives a beautiful appearance to the galleries. Sometimes when a hewer detaches a block of ore with his pick mass of quicksilver, the size of a pigeon's egg, rolls out, and leaping along the floor, divides into thousands of small drops. Owing to the imperfect apparatus with which the ore is sublimated, nearly one-half is lost. Formerly criminals only were employed in these mines. They were conducted at sunrise from prison by a subterranean passage into the mine, and compelled to toil on until the evening, when they were led back again to their dungeons. In a few years the greater number died, through inhaling the poisonous vapours of the mercury. Reduced to despair, a century and a-half ago, they set fire to the galleries, which, being then constructed of wood, were destroyed, and mining operations put a stop to for many years. Only free labourers are now employed, who are not allowed to work longer than six hours a-day. Most of these, however, die between the ages of thirty and forty, and those who exist longer are affected by palsy.

The quicksilver mines of Idria were discovered upwards of three centuries ago by a peasant who had placed a tub under a spring issuing from the mountain side. On attempting to move it, he found it excessively heavy, and on examining the bottom he saw that it was partly full of a heavy liquid, shining like silver. Ignorant of the value of the substance, he had sense enough to take it to a goldsmith, without mentioning the place where he had found it. In course of time, however, a man named Anderlein, having bribed him, became master of the secret, and with several others began to work the mine. In the next century the Venetians drove out the Germans, but were finally compelled by the Emperor Maximilian to give it up, and he restored it to its rightful owners. The mine has since been worked by the State. Ingress to the mine can be obtained by descending a convenient flight of steps, with galleries running off here and there from landing-places, or by descending in a few minutes through a perpendicular shaft in one of the tubs by which the ore is raised. The galleries lead to the various storeys of the mine, the lowest of which is 145 fathoms beneath the surface. The vein itself descends to an unknown depth, and is horizontal, but its extent has not yet been measured. The ores being embedded in limestone of a loose nature, all the galleries had from the first to be supported by wooden props. The wood has, on several occasions caught fire, with disastrous results. Early in this century the labourers observed a thick smoke issuing from the deepest part of the mine. It rose higher and higher, spreading through the upper galleries, yet no fire was to be seen, nor sound of flames heard.

Some of the workmen attempted to reach the scene of the fire, but were driven back by the dense and suffocating smoke, impregnated with vapours. Endeavours were made to smother the fire, but though the mine remained closed for five weeks, no sooner was it re-opened than the fire burst forth more furiously than at first. The howling of the flames ascending from the lowest depths of the pit awed the spectators, and the mercurial and sulphureous fumes arising from it threatened instant destruction to all who might approach. The director of the mine, as a last resource, came to the decision of flooding the works, and a river turned into the shaft ran down it for two days and three nights. At first no perceptible effect was produced, but on the second a terrific explosion shook the mountain as if an earthquake had taken place. The huts near the opening were blown to pieces, and even the stone houses on the slopes of the hill, fell with tremendous crashes. Water, however, gained the victory. Gradually the vapour dispersed, and after a few weeks the workmen were able to descend into the pit. They found, however, the galleries torn up, the vaulted roofs burst, and the stairs destroyed. It took two years to pump out the water, which, it is said, poisoned all the fish in the Idriza.

High pay being offered to any who would venture in to collect the quicksilver, which had accumulated in considerable quantities, many, tempted by the bribe, made their way into the workings, but overcome by the mercurial vapours, several perished.

The galleries have now been formed of stone, seven feet high and six feet broad, though some are still propped up with wood. They are of immense extent, amounting to no less than fifty miles. As late as 1846 another fire occurred in the wooden galleries, which was quenched by putting that part under water. The workmen labour in a tropical heat and an atmosphere full of deadly vapours. It is no wonder that a premature age overtakes many of them, and that young men are seen trembling in every limb, though it is said that those who survive their forty-fifth year may live on until they are sixty or seventy. To transport mercury, the greatest care is required. It is first packed in sacks of sheepskin, tanned with alum. The sack, being pressed and punched to ascertain if it is sound, is enclosed in a second skin. These are then placed in a small cask, and the cask again in a square box. Notwithstanding these precautions, as the sacks sometimes burst, the loss of the metal is great, and the mercury is now generally transported in large iron bottles, the stoppers being screwed down by means of a machine; in this condition, it is exported to England.

Chapter Seven
Stalactite and Ice-Caverns

Numberless and varied are the cavernous regions below the earth, presenting the strangest and often awe-inspiring sights to the spectator. In some rivers flow hundreds of feet beneath the green fields and widespreading trees.

Through the caverns of Adolsberg, Planina, and Upper Laibach flows a river known as the Poik, which then assumes other denominations, according to its locality. In some places it forms cataracts, leaping over the most picturesquely grouped rocks. In others it has forced a passage amid them, and then flows gently on.

Our travellers resolved to undertake a voyage on the Poik, and embarked in a boat, their progress being stream upwards through the celebrated cave of Planina. They had to be cautious, for often the current ran with great rapidity, and to keep a watchful eye for rocks which lay hidden beneath the water.

Rowing on for about 600 feet from the entrance of the cavern, at the end of a magnificent dome, they found that the river occupied the whole space. To this part persons on foot could proceed, as the ground on either side of the river was level. Now passing through a portal 48 feet high and about 24 broad, and as well proportioned as if cut out by the hand of man, their ears ere saluted by the thundering roar of a distant cataract. As the archway idened, they suddenly emerged on a lake 250 feet in length and 150 broad, eyond which the cave divided into two arms, forming the channels of two reams, whose confluent waters formed the lake.

The walls of the cavern on either side rose abruptly out of the water, th the exception of one small landing-place at the foot of a projecting lge. Here and there hung masses of stalactite, resembling a petrified scade, the rest of the rock being black and naked. So high was the vault iat their torches could not pierce the gloom, the impressiveness of which as increased by the roar of a waterfall heard through the channel to the t.

Hitherto their progress had been easy, but they now resolved to proceed up the left branch. They had frequently to get out of their boat, and wading, drag her over the shallows. The voyage terminated at the end of a small hall with a circular dome, the floor being a lake 180 feet in length, and from 40 to 45 feet in depth. In the roof appeared a chasm, sloping upwards through a small aperture, in which a violent current of air set in, almost extinguishing their torches.

Beyond the mouth of the chasm another gallery opened out, into which the persevering travellers penetrated. Nothing could surpass the beauty of the spar crystals with which its walls were encrusted. At the entrance stood a white figure, which might easily be supposed to be an angel, guarding the entrance with a glittering sword, threatening all who should venture with profane hands into his sanctuary.

Further on, projected in bold relief, was a colossal statue of a monarch, sceptre in hand. As they proceeded they passed groups of stalagmitic cones of all shapes and sizes. Some like the smallest icicles, others rising six feet in height from the ground, as thick as a human figure, the whole shining and glittering as the light of the torches fell upon them, and standing out in bold relief against the dark background formed by the brown wall of the cavern.

Returning to the central hall, they made their way up the eastern branch, which is much larger than the one they had just visited, the main stream flowing through it. As they pulled up, the increasing roar of waters announced a large waterfall. They found that enormous masses of stone, falling from the roof, had narrowed the bed of the river to about fifteen feet, over which the water shot in a broad sheet, fully ten feet in height. The effect as it rushed over the jet-black rocks, casting up flakes of milky white foam, when illuminated by the torches, was very beautiful.

Having hauled up the boat over the rugged mound, they again embarked, encountering a couple of reefs. They then proceeded on between steep walls with a free navigation, for upwards of four miles. In many places the roof was adorned with draperies formed of snow-white stalactites, but generally the black walls alone appeared. In some parts the roof descended so low that they were compelled to lie down, and shove the boat along by holding to the roof above their heads, until at length they found that they could proceed no further.

Of the world beneath the surface some of the most beautiful scenes are presented by the ice-caves of France and Switzerland. One of the most curious is the glacière "Grâce Dieu," near Besançon. In the centre of the cave rose three stalagmites of ice. The central mass was 66.5 feet in circumference. Some distance above the ice-floor on the right was a small fir-tree, which had been fixed in the ground, and had become completely covered so that the tree itself had disappeared, its crystal incrustation showing every elegance of variety in form. From each twig of the different boughs, complicated groups of icicles streamed down. The mass to the left, however, was the grandest and most beautiful. It consisted of two vast heads, with several others of less height resembling a group of lions' heads bending down, richly decked with icy manes, huge masses measuring 76.5 feet in circumference. On looking at this column from the side opposite the entrance to the cave, so that it stood in the centre of the light pouring down in a long slope from the outer world, the transparency of the ice made the whole appear as if it were set in a frame of impalpable liquid blue, the effect of the light penetrating through the mass at its extreme edges.

The Schafloch or Trou-aux-Moutons, a vast ice-cave on the Rothhorn, in the canton of Berne, is equally beautiful and curious. It takes its name from the fact that on the approach of a storm, the sheep and goats fly to it for shelter, although never going as far in as the place where the ice commences. The travellers entered the cave amid masses of loose stone, with which in a short time the ice was found to intermingle until it entirely hid the naked rock. They passed between two magnificent columns of ice which formed the portal to the fairy cavern. The floor, composed of ice, rose on either side to meet these columns in a graceful swelling curve, so that it appeared as if their bases expanded and met in the middle of the cave.

They had now to make their way amidst stalagmites rising from the floor, met by stalactites descending from the roof. All the time as they twisted in and out among the glittering pillars of ice, endeavouring to do as little harm as possible, they were accompanied by an incessant fall of small portions, shivering and glittering on reaching the ground.

Passing beyond the two columns, they saw before them a perfect sea of ice, which became broader and broader until they reached the edge of a magnificent ice-fall, smooth and unbroken, beyond which they were unable to penetrate.

They afterwards visited another beautiful ice-cavern known as the glacier of Saint Sivres, into which a stream flows, becoming completely congealed.

There are many other ice-caverns in Bohemia, Hungary, the Hartz Mountains, and in various parts of North America. One of them, however, surpassing in size the others, is the cave of Yermalik, in the province of Kondooz, in the centre of Asia. When Kondooz was invaded by the savage warrior Genghis Khan, 700 men with their wives and children took refuge in this cavern, and offered so brave a defence, that after attempting in vain to destroy them by fire, the barbarous invader built up the entrance with large blocks of stone, and left them to perish of hunger.

Nearly forty years ago the cave was visited by two British officers, who had great difficulty in obtaining guides, as the natives believed the cave to be the abode of Satan. The entrance is about half-way up a hill, and about fifty feet in height, and about the same in breadth. Squeezing their way through a narrow passage between two rocks, probably the remains of Genghis Khan's fatal wall, they came to a drop of about sixteen feet. Down this, by means of ropes, they were lowered by two men, who remained to haul them up again. Passing through a narrow tunnel, over a floor of smooth ice, they reached a vast hall, damp and dripping, the light of their torches not enabling them to form any idea of its size. Here they discovered hundreds of skeletons, the victims of Genghis Khan's cruelty. Among them was one, evidently the skeleton of a mother, holding in its long arms the skeletons of two infants. The bodies of others had been preserved, and lay as they had fallen, shrivelled into mummies. After leaving this vast sepulchre, they proceeded through several low arches with smaller caverns, until they reached an enormous hall, in the centre of which was a prodigious mass of clear ice, in the form of a bee-hive, its dome-shaped top just touching the long icicles which depended from the jagged roof.

A small opening led into the centre of this wonderful ice-heap, which was divided into several compartments, presenting numerous fantastic forms. In some the glittering icicles hung like curtains from the roof, in others the whole compartment was as smooth as glass. The prismatic colours which presented themselves as the torches flashed on the surface of the ice were beautifully brilliant.

On every side they were surrounded by solid ice, and, scarcely able to keep their feet, they slid noiselessly over the glittering surface of the mysterious hall.

The icicles having reached the floor of one of the largest of the compartments, had the appearance of pillars supporting the roof.

In Italy and the South of France there are caverns with some distant aperture through which the wind enters, and being cooled in its subterraneous passage, sends forth a cold blast at the other end, such as the Aeolian Cavern, near Terni. It has been utilised by the proprietors of some of the neighbouring villages, who have conducted the cold air to their houses by means of leaden pipes, which on sultry summer days convey a pleasant coolness through plaster-of-paris masks, with wide distended mouths, fixed in the walls of the rooms.

Chapter Eight
Copper Mines

Next to England, Sweden is one of the chief copper-producing countries of Europe.

The mine of Fahlun, in Dalecarlia, has been worked from times immemorial. In consequence of the careless way in which the excavations were propped up, in the year 1678 the surface of the ground fell in, forming a vast pit of above 180 feet in depth, 1200 feet long, and 600 feet broad, with precipitous and sometimes overhanging walls, so that the spectator appears to be standing on the brink of an enormous crater. The bottom is filled with masses of rubbish and the remnants of ancient shafts, and thick beams of wood are seen protruding in all directions. A broad and convenient wooden staircase has, however, been formed on the northern side of the pit, by which not only the miners, but even horses can descend to their work. Passing through the entrance, the mine gradually widens underground to a depth of 1062 feet. The chief mass of ore is 600 feet broad on its upper surface, greatly narrowing as it descends to a depth of 1200 feet. Round it are other similar deposits. As the copper pyrites are deposited generally on the circumference of the outer shell of these masses, which are of a very irregular outline, the mining operations are carried on in a perfect labyrinth of winding passages and galleries, situated at various depths, and supported either by pillars or walls. It at one time yielded 5000 tons of copper annually, but has of late years furnished no more than 600 tons.

A romantic incident is connected with this mine. In the year 1719, while some miners were exploring an abandoned passage, they discovered a human body, preserved from corruption by the blue vitriol or sulphate of copper produced in the mine under the influence of the atmosphere and water. It was that of a handsome young man. On being brought to the surface, people from all directions flocked to see it, but nobody could recognise in its features a lost kinsman or friend. At length a woman, with tottering steps, upwards of eighty years of age, approached the corpse, when scarcely had she cast a glance at it than she uttered a piercing shriek, and exclaimed,—"It is he! It is Gustavus, for whom I have mourned so long, whom I accused of fickleness in deserting me."

She had in truth recognised her affianced lover, who had mysteriously disappeared more than sixty years previously, but whose image she still bore in her memory. As he was not employed in the mines, no one thought of searching for him underground. The surface is traversed by various crevices, some leading to the workings underground; and probably Gustavus, prompted by curiosity, had looked down one of them, and had either, losing his balance, fallen in, or been precipitated by some jealous rival in the good graces of the once blooming girl, now a tottering old woman, weighed down with a double burden of infirmity and age. She probably forgot how years had passed away, as she gazed once more on the face of her youthful and handsome lover.

Besides copper, Sweden produces iron of great excellence, won from its celebrated mines of Dannemora, and largely imported into England for the manufacture of steel.

Leaving the university town of Upsala, and passing through a natural barrier of forests and lakes, in which lie the iron-works of Oesterby, the travellers reached the place in which the pit of Dannemora is situated; not a sign announced the vicinity of the mine, until they saw the machines for lifting the ore, and a few huts scattered about, when they found themselves standing on the brink of a vast pit or crater, whose black and precipitous walls fence an abyss of a mile in circumference, and a depth of 450 feet. Here and there in that cold region they perceived patches of perennial snow and along the black walls, the dark entrance to labyrinthine caves fringed with long stalactites of ice. In some of these hollows flames were seen creeping along the cliff as they issued from piles of fir wood to soften the hard rock, while on every part of the deep gulf human beings were at work, the clang of their hammers sounding like the clicking of numberless clocks, mingled with the creaking of machinery, which brings to the surface the casks of ore. At length a bell tolled, and men, women, and children were seen ascending in the tubs, some standing on the edges, holding on with perfect confidence to the rope by which they were hoisted up.

Silence now reigned below, except when the voices of overseers were heard summoning those who had lagged behind, to ascend in haste. Scarcely had they reached the upper surface when a loud thundering roar was heard, which echoed through the cavern. The ground trembled as if convulsed by an earthquake, while black masses of smoke with pieces of stone or ore ascended from the gulf, and the crashing sound of falling masses rent from the mother earth was heard. When all the charges had exploded, the miners again descended to their work.

Although it cannot be classed among the wonders of the subterranean world, the famous Erzgebirge or iron mountain in the Styrian Alps deserves mention. It rises to the height of 3000 feet, the whole being coated with a thin mantle of the richest ore. In all directions it is covered with machines of various forms, horizontal and vertical galleries, tunnels and roads, and represents, as it were, a mine turned inside out. The whole of the operations are exposed to view, like those in the Carclaze tin mine in Cornwall, only in the former the ore is conveyed by tram-roads, galleries, and shafts to the bottom of the mountain, where they all unite in one main shaft, from which a tramway runs to the smelting-ovens of Eisenerz and Vordernberg.

Among the beautiful productions of nature, rock-crystal may be classed, known as the false topaz when yellow, the morion when black, and the smoky quartz when brown. The colourless kinds are often called Bristol or Irish diamonds, and the violet the amethyst. Some few years ago, a party of tourists, led by a guide, Peter Sulzer, set out from Guttannew, in Switzerland. When descending the mountain they reached a dark cavity, out of which they extracted some pieces of black rock-crystal with the handles of their Alpine stocks. The following year, Sulzer and his son, with a few companions, made an attempt to force their way into the cave, by widening the entrance with gunpowder. In spite of hail, rain, and bitter cold, they persevered, remaining during the night close to the cavern, in order to renew their labours the next morning.

Having widened the entrance, they penetrated to a considerable depth into the mountain, through a large cave piled up with débris, in which were embedded large planes of jet-black morions. These beautiful crystals had grown originally from the sides or roof, and had either fallen from their own weight, or been shaken out by some convulsion of nature. Their toil was rewarded by upwards of a thousand large crystals, varying from fifty pounds to more than three hundredweight.

Their expedition and its result becoming known, the whole population of Guttannew turned out with hammers, spades, and baskets, to carry off what they had left. As it was reported that the Government intended to interfere, they laboured night and day for a week, until, by the time the authorities arrived from Ijri, the whole had been removed. Some of the finest specimens are still to be seen in the museum at Berne.

Amber, about which all sorts of fabulous stories have passed current, is found more frequently in the depths of the sea than in those of the earth. There can be no doubt that it is the product of several conifers, or cone-bearing trees, overwhelmed by the waves. Although the gum which exuded from them has remained concealed for ages, until washed up from the bottom of the ocean, flies and spiders, which must have been caught when it was in a semi-fluid state, have been found embedded in it. The insects now appear as perfect as they were thousands of years ago.

The naturalist, Dr Berndt, has discovered 800 different species of insects in amber.

The famed cavern in Kentucky is as well worthy of a visit as any subterranean region. Of late years an hotel has been built near the entrance, detracting from its once romantic appearance. Visitors first descend a well-like pit, into which a stream falls, by a flight of steps, and then passing under a high archway, proceed along a level road, to what are called the vats, where saltpetre was once manufactured. Their blazing torches, numerous as they may be, hardly light up the vast subterranean region. From the large hall they make their way through a low narrow passage, known as the "Vale of Humility," into another hall of enormous extent, the roof so lofty that the torches scarcely illuminate either the walls or roof. At their feet can be seen the glitter of water, extending far away into the interior, a bright stream flowing over a rocky bed into it. Moving on, they in a short time reach Echo River, on the shore of which a boat is found. When looking upwards, it appears as if a canopy of black clouds hung over their heads. On either side can be seen precipitous cliffs, rising apparently into the sky. Silence and darkness reign around, the smooth sluggish water alone reflecting the glare of the torches. The visitors are not disposed to utter a word, until the voice of one of the native guides suddenly bursts forth into a melancholy chant, which seems as if echoed by the spirit of his departed brethren. Now the notes rise, now they fall, as he gives them forth with the full force of his lungs, or warbles softly, finishing with a melancholy wail, which produces a most mournful effect. When a pistol is fired off, there comes a succession of crashing thundering sounds, echoed from every angle of this enormous vault; backwards and forwards they rush, roaring and reverberating from wall to wall with terrific crashes. The guides say it is perfectly safe at all times of the year to traverse the cavern, but there have been occasions when the waters, rising suddenly, have prevented the return of explorers. A way, however, was at length discovered through a narrow passage, the course evidently, at one time, of a stream, up which they can climb over the mud, and save themselves from being drowned or starved. This passage has appropriately been called "Purgatory." In one part the river expands into a lake, the gloomy effect of whose dark waters, lost in the darkness, is indescribable. Leaving Echo River, they enter another cavern, known as Cleveland Cabin—a fairy region. Above their heads, and on either side, the roof and walls are adorned with delicate flowers, of snowy whiteness, and domes, turrets, spires, shrubs, and trees, as well as with the forms of birds and beasts of all descriptions; indeed, figures of every shape which imagination, without any great exertion, can picture, appear around. The representations of some are so perfect, that it is difficult to believe that they have not been carved by the hand of man, and yet all of them have been produced by the dripping of water from the gypsum rock. The cavern is not destitute of inhabitants. Huge crickets and spiders of an almost white colour crawl along over the ground, and rats as big as leverets run by, exhibiting sharp teeth and long tails. Another cavern is called "Martha's Vineyard." It appears as if a vine had climbed up the sides and spread its branches over the roof, from which hang suspended what look like clusters of grapes, but all of the same stony nature. In another cave it seems to the visitor that he is standing in a wintry scene, ice above and ice on the ground, with here and there patches of snow, the appearance being caused by the excessive whiteness of the gypsum. Farther on, there is a beautiful grotto, called "Serena's Arbour," the walls of which are covered with a drapery resembling yellow satin, falling in graceful folds, while through it murmurs a rivulet, which makes its way to one of the many rivers running through the cavern. In another, on the torches being extinguished it appears as if stars innumerable were glittering in the sky. On a stone being thrown upwards, it quickly strikes the roof, and it is soon seen that these seeming stars are produced by pieces of mica embedded in the roof, on which the light of a lantern being thrown in a peculiar way is brightly reflected. Although the caverns seem to be of immense height, the ceiling in most parts is not more than thirty feet from the ground. In the centre of one cavern, a regular hill rises from the ground, with a stream running at its base. Several rivers are crossed in this vast cavern, one is called the Echo River, another the Styx, and a third the Lethe. They are inhabited by fish and crawfish, sightless and perfectly white.

This vast cavern, the ramifications of which are said to measure nine miles, was not known to white men until 1802. For many years no one advanced beyond three miles from the entrance, further progress being stopped by a deep cavern called the "Bottomless Pit," 1000 feet deep. At length, however, a daring guide threw a ladder over it, and crossing by this means, he was able to explore six more miles of this subterranean region.

A bridge has now been constructed by which people can pass over in perfect safety. It is said that no dog will willingly enter the cavern; indeed, few persons can pass along its passages without a sensation of awe, although with a guide it may be traversed without danger.

Chapter Nine
Silver Mines, etcetera

If a true history of the silver mines of South America were to be written it would reveal the cruel death of thousands and thousands of human beings, sacrificed to the lust of gain. High up among the Andes, surrounded by a succession of steep and naked rocks, is the town of Pasco, built above the mines, from which the inhabitants obtain their subsistence. The entrances to most of the mines are situated in the midst of the town. The irregular shafts descend directly down into the interior of the mountain, access being by a series of ladders often ill-constructed and rough, ropes and chains being employed to hoist up the ore. Frequently, the overseers having neglected to put up the necessary props, portions of the mines have broken in and destroyed many of the hapless workers. In one instance 300 perished at once by this means. In most of the mines the labourers, after getting out the ore, have to bring it to the surface in baskets on their backs, often from immense depths, and were it not for the sustaining coca leaf they would be unable to undergo such excessive toil. When rich veins are struck, the wages of the miners increase, but in most instances they spend them in drinking and debauchery, while the proprietors of the mines are almost equally uncivilised.

Fourteen miles from the town of Caxamarca is an isolated mountain called the Cerro de San Fernando de Gualgayoc, traversed by numberless veins of silver. At its summit rise a number of pyramidal pinnacles. Its steep sides are pierced by several hundred galleries formed for the extraction of the ore, as well as by numerous natural openings, while in all directions are seen the huts of the labourers, sticking like the nests of birds, wherever a ledge has enabled them to be constructed. One of the richest silver mines of Peru is that of Salcedo, but nothing is now known of it except its tragical history. A Don Jose Salcedo, a Spaniard, without a maravedi in his pocket, made love to an Indian girl, whose mother promised to reveal to him a rich silver lode on condition that he married her daughter.

Aided by his Indian relatives, with whom he lived on the most friendly terms, he obtained vast quantities of silver from the mine, the entrance to which was kept carefully concealed.

His wealth excited the rapacity of the viceroy Count Lemos, who, to obtain possession of it, accused him of exciting the natives to rebellion, and cast him into prison. In vain Salcedo entreated that he might appeal to the mercy of the king, and promised to give the viceroy a bar of silver daily, from the time the ship left the port of Callao to her return from Europe, which would probably be upwards of a year; but the viceroy, instead of listening to the proposal of Salcedo, ordered him to be hung. No sooner was this known to the natives than they destroyed the works, and so carefully concealed the entrance, that even to the present day it is unknown. The tribes afterwards dispersed, and even cruel tortures could not induce them to reveal the secret.

There can be no doubt that there are many rich lodes in existence worked by Indians, who, knowing that they will be compelled to labour for the benefit of their masters, carefully conceal them. In many of the mines of Peru, the natives having almost been exterminated, the proprietors endeavoured to kidnap the inhabitants of the Pacific to supply their places, but after several hundreds had been nefariously captured, the Governments of England and France interfered and put a stop to the practice. In another part of South America, near the town of Cumana, is a vast cavern in the Valley of Caripe, which was many years ago visited by Baron Humboldt, who found it inhabited by a remarkable species of nocturnal bird, called the guacharo. The mouth of the cavern is pierced in the side of the cliff looking towards the south, in the form of an arch, eighty feet wide and seventy-two in height. The summit of the cliff is covered with trees of gigantic size, and with shrubs and plants growing in all the luxuriance of a tropical vegetation, while a variety of creeping plants hang in elegant festoons before its entrance. Visitors can proceed for upwards of 430 feet without being compelled to light their torches. When the light of day begins to fail, the hoarse cries of the nocturnal birds are heard coming out of the dark recesses of the interior. The guacharo is of the size of the common fowl; its hooked bill is white, like that of the goat-sucker, and furnished at the base with stiff hairs, directed forwards. The plumage is of a sombre brownish grey, mixed with black stripes and large white spots. Their eyes are incapable of bearing the light of day, and their wings are disproportionately large, measuring no less than four and a-half feet from tip to tip. The birds quit the cavern only at nightfall, to feed on fruits. A most horrible noise is made by them in the dark recesses of the cavern, and the clamour increases as they are disturbed by the visitors advancing deeper into it with torches, and those nestling in the side avenues begin to utter their mournful cries. When the first sink into silence, it seems as if the more remote inhabitants were alternately complaining to each other of the intruders. The nests of these birds are fixed fifty or sixty feet from the ground, in funnel-shaped holes, with which the cavern roof is pierced like a sieve.

Armed with poles, the natives once a year, about mid-summer, enter the cavern and knock down the young birds, while the old ones, with lamentable cries, hover over the heads of the robbers. The young which are taken are opened on the spot, when the peritonaeum is found loaded with fat, and a layer of substance reaches from the abdomen to the vent, forming a kind of cushion between the bird's legs. At this period, called by the Indians the oil harvest, huts are erected by them, with palm leaves, near the entrance. Here the fat of the young birds is melted in clay pots, over a brushwood fire; but although thousands are killed, not more than 160 jars of clear oil are obtained. A small river flows through the cavern, and the visitor is compelled, as he proceeds, to wade through water, not, however, more than two feet deep. From the entrance as far as 1458 feet the cavern maintains the same direction, width, and height, after which it loses its regularity, and its walls are covered with stalactites. The same bird has been found in the province of Bogota, and may probably be discovered in other caverns. Animal life exists in considerable quantities in many subterranean regions, such as beetles, eyeless spiders, scorpions, millipedes, and crustaçeans. The most curious is the Proteus anguinus, which breathes at the same time through lungs and gills. It has a long eel-like body, with an elongated head, and four very short and thin legs. The skin is flesh-coloured, and so translucent that the liver and heart, which beat about fifty times a minute, can be seen distinctly beneath. Two little black spots, resembling eyes, lie buried under the skin, and are only partially developed. Weak as it appears, it glides rapidly through the water, when its four little legs remain motionless; it uses them, indeed, only for creeping, and then in a very imperfect manner. Seven distinct species of proteus have been discovered, six of which were found in the cavern of Carniola, besides crickets, spiders, and a few crustaceae. A peculiar blind rat is found in the Mammoth Cave of Kentucky. A blind fish swims in its rivers, and Professor Agassiz is of opinion that they, like all other blind animals of the cavern world, have at no time been connected with the world of light.

Vegetable life also exists in caverns, but consists of such mushrooms or fungi which, shunning the light, love darkness and damp. For their existence, however, moisture and warmth of air is necessary, but they are invariably dependent on organic basis, and are commonly found germinating on pieces of wood, particularly in a state of decomposition. More than seventy subterranean fungi have been discovered, some remarkable for their size. A few years ago a fungus was found growing from the wood-work of a tunnel near Doncaster, which measured no less than fifteen feet in diameter.

In the neighbourhood of Paris the cultivation of edible mushrooms is extensively carried on in the catacombs or caverns, seventy or eighty feet below the surface, where the temperature is uniform all the year round. In one of the caves of Mount Rouge there are no less than six or seven miles of mushroom bedding. Among the wonders of the subterranean world must be classed the bone caves of Europe and other parts of the world. In some caves in England, the bones of a prodigious bear have been found, and many hundreds of those of a hyena, considerably larger and more formidable than those existing in Africa. Besides the bear and hyena, upwards of a hundred species of extinct animals have been found in the ossiferous caves of Great Britain, among them being those of the elephant and a rhinoceros. Though in Europe bone caves contain the remains of animals very different from those now existing in the same regions, yet in the caves of Brazil extinct species of nearly all the territorial quadrupeds now inhabiting this region occur. The Australian caverns contain fossil bones of a large extinct kangaroo. In New Zealand the wingless apteryx is still found in the wilds, and the caves of that country show us that it was preceded by other wingless birds of gigantic stature; among them the moa, which, when alive, must have stood about thirteen or fourteen feet high. A complete leg of the bird has been discovered six feet in length, and portions of the eggs show that they had been about 6 or 7 inches diameter.

Chapter Ten
Arrangements of the Mines

In Germany mining operations are carried on in the most systematic manner. Miners are dressed as their ancestors were hundreds of years ago, and they cling pertinaciously to their ancient usages. In some workings prayers are offered up, led by the engineer, before the miners descend to their work, while they stand grouped round him at the opening of the mine, a custom which might well be adopted in our own country. The German miner retains also the superstitions of his forefathers, and still believes in the genii of the mines, named Nickel and Kobald, after whom he has called two metals, nickel and cobalt, originally discovered in the mines of Saxony.

The Germans have introduced into their mines a regular military system, and the engineers, who are denominated captains, wear when in full dress a uniform of a very military appearance, set off by epaulets and gold embroidery. Not inferior to them, however, are the Cornish miners, their captains being those who have risen by their industry and intelligence from the lowest to the highest grades, although men of less education than their German brethren.

The Spanish miners are a sober and frugal race, enjoying their cigarettes even while at work. On leaving the mine they put on their snuff-coloured cloaks and broad-brimmed sombreros. In the southern part of the Peninsula they wear grass sandals, cloaks of bright colours, and handkerchiefs bound round their heads. Leading lives of toil and hardship, their huts are wretched abodes built of stones and mud, their beds the ground, an iron or copper kettle hung from the roof above the fire in the centre of the cabin, a few wicker baskets, and a waterbottle of porous clay constitute their furniture. Still, the lot of the miner of the Sierra Morena is far superior to that of the miner of Almaden, who, poisoned by the noxious vapours of mercury, quickly succumbs, ere he has gained the prime of manhood.

In South America the mining operations of the inhabitants somewhat resemble those of their Spanish ancestors, their habits and customs being imitated by the Indians, who have, however, to perform the harder part of the work. While Mexico and Peru were under the mother country, the Mita or law of compulsion existed, the Indians being forced to toil against their will in the mines, but since the emancipation of the colonies and the abolition of that nefarious law, they have returned to their agricultural pursuits, and are only occasionally found of their own free will labouring in the mines.

Various modes are adopted for descending the mines. In some merely a single rope or chain with a loop at the end in which the miner places his foot is used, even when the depth is several hundred feet; in other mines baskets or tubs in which three or four men can stand are employed. While one of these is hauled up, another descends, and often fearful accidents have occurred by the tubs striking against each other, when their occupants have been thrown out. Occasionally the ropes and chains have given way, and the hapless miners have been dashed to pieces.

Some few years ago, as the engineer and several men of the mine of Méons were descending standing in a tub, each with a lamp in one hand, and holding on to the chain above him with the other, a couple of tubs loaded with coal unhooked theirs, which fell to the bottom. Providentially they had not relaxed their grasp of the chain above their heads, and at once letting go their lamps and desperately seizing it with both hands, they continued their descent, though huge lumps of coal were falling out of the tubs above them. Wonderful to relate, they reached the bottom in safety. On another occasion, while the same engineer was ascending in a tub, it was upset in consequence of the engineman raising the rope too suddenly. The engineer hanging on by one leg, with his head downwards was hoisted a height of forty yards, before the alarm was given and he was lowered to the bottom.

In the same mine, another engineer, while descending in a tub, had his clothes caught by a strut which projected from the side of the pit; he here hung suspended while his companions continued to descend, terrified for his safety and alarmed for their own, as should he fall, they expected to be crushed by his weight. In vain they shouted for assistance, the men at the top of the pit having gone out of hearing. Not until they reached the bottom could they send any aid to their companion. He in the meantime had been vainly endeavouring to find some support so as to relieve the strain on his torn garments, which threatened every instant to give way. After hanging thus for twenty minutes, he was at length set free, but no sooner was he received in the tub than he became insensible. A severe illness of long duration followed, but he ultimately recovered, though he ever afterwards preferred going down the ladders to descending in a tub. Anecdotes of the same description could be given without end. Most accidents of this character have ended fatally. To avoid them various inventions have bee devised, one of which is known as the mounting machine, or man-engine. consists of two parallel rods, furnished at equal distances with steps, whi one is raised to a certain height the other is lowered to the same distanc While the movement of the crank is on its turning point, the miner pass from the step on which he is standing to the opposite step of the other. A they are constantly moving up and down, his next step is back again to th rod he had before left, which rising a few feet, he is able to step back to th other, just as it, having gone down, is once more ascending; and thus h reaches the top with little fatigue.

Far superior to this mode of ascending or descending are the safet cages introduced of late years, which have guides the whole length of th shaft, and bonnets or roofs to protect the heads of the men within. The are made with several stages, in which either the tubs or waggons can b placed, or where the miners can stand or sit. If a rope breaks, a spring place above the cage and kept taut by the tension of the rope, is set free, and ac upon a double clutch made of the best tempered steel. This catch or wedg falls between the wooden guide and a part of the cage, and brings the latt immediately to a stand-still. By this means numberless accidents have bee prevented. The man-engines which have been described are dangerous f novices, for should a person stop at the wrong time, he may be hurled to th bottom, or crushed at the return stroke.

One of the most frequent accidents to which miners are exposed aris from an outbreak of fire-damp. To avoid this, various safety-lamps hav been invented. The most celebrated is that known as Sir Humphrey Davy lamp. The flame is enclosed in a fine wire gauze, through which, und ordinary circumstances, the gas cannot penetrate. There are other lamps use constructed on the same principle, but superior in some respects. To often, however, the miners open them at some fatal moment, or enter th mine, against orders, with naked candles. Still, by means of these lamps when properly employed, many accidents have been prevented. Anothe invention exists by which a person can enter in the midst of impure ai The apparatus was devised by Monsieur Kouquayrol, a French engineer. consists of a reservoir made of sheet iron, into which the air is forced, anc by an ingeniously contrived pump, is secured like a knapsack to a man' back, and the air is conveyed by means of a tube to the mouth of a nose, an thus into the lungs at the ordinary pressure, while a small external valv allows of the escape of the air after it has been respired.

A still more simple apparatus has been invented by Monsieur Galibert. The system for condensing the pure air is more perfect, while the reservoir consists of a well-prepared goat-skin, which, when inflated, a man can with ease carry on his back. It is furnished with a similar contrivance to the former, a tube passing from the reservoir to the mouth, while the nostrils are compressed, the eyes and head are protected, so that provided with it, a person may exist for a quarter of an hour in the foulest atmosphere, or in the midst of dense smoke. Although the metal miner is subjected to fewer accidents than are his brethren working in coal mines, the atmosphere in the former is far more destructive to human life. In lead mines, the duration of life averages scarcely more than thirty-two years, and in those containing arsenical pyrites or quicksilver ores, the average is still lower. Before the use of gunpowder in underground operations, the rocks containing the ore were attacked with fire, indeed the practice is still retained in some countries. Huge wood fires are made up against the face of the rock, which becomes shattered and traversed by cracks, and when cooled, it is easily detached with a pick or fork. Of late years, however, machines have been devised for boring or breaking the rock. Some form a hole by the continuous motion of a rotating drill, others by means of intermittent blows. One of these rock-boring machines, manufactured by Messrs Turner, of Ipswich, performs its work by a combination of both these operations. By the employment of these machines, the formation of the tunnel under Mount Cenis was greatly facilitated. An example has already been given of the way in which people have been saved from the effects of inundations in mines, others have been dug out when buried by the fall of roofs, but almost countless are the numbers who have perished from other causes, for if the first have destroyed their hundreds, the fire-damp in coal mines has proved the destruction of thousands. It was at one time considered right every night to provoke an explosion by lighting the fire-damp in order that the working stalls should be accessible next morning. The man who performed this dangerous operation wore a thick covering of wool or leather, his face was protected, and his head was covered by a hood like a monk's cowl. He crept along the ground, carrying in his hand a long pole with a light at the end of it. He was known in the English mines as the fireman, but in the French he was called either the cannonier, the monk, or the penitent, the latter name being given him from his dress resembling that of certain so-called religious orders in the Romish Church. Too frequently the hapless penitent was destroyed by the explosion he had provoked.

Our two friends, however, might have written several large volumes had they given accounts of even a portion of the interesting matters concerning mines which they gathered up during their long and varied tour.

Mark did not fail to benefit largely by the information he obtained, and he ultimately, with the numerous improvements he introduced, became the proprietor of two of the coal mines in which he had worked in his boyhood, while his young sister, on whom he had had the satisfaction of bestowing a high-class education, refined in mind and manners, became the wife of his friend and fellow-traveller.